How the Earth Works
Part IV

Professor Michael E. Wysession

THE TEACHING COMPANY ®

PUBLISHED BY:

THE TEACHING COMPANY
4151 Lafayette Center Drive, Suite 100
Chantilly, Virginia 20151-1232
1-800-TEACH-12
Fax—703-378-3819
www.teach12.com

ISBN 1-59803-407-3

Michael E. Wysession, Ph.D.

Professor of Geophysics
Washington University in St. Louis

Michael E. Wysession is Professor of Geophysics at Washington University in St. Louis. He earned his Sc.B. in Geophysics from Brown University and his Ph.D. from Northwestern University.

Professor Wysession has established himself as a world leader in the areas of seismology and geophysical education. He has developed several means of using the seismic waves from earthquakes to "see" into the earth and create three-dimensional pictures of Earth's interior. These images help us to understand what Earth is made of and how it evolves over time. An important focus of Professor Wysession's research has been the complex boundary region between the solid rock of Earth's mantle and the liquid iron of Earth's core. Another focus has been the identification of large regions of water-saturated rock in the deep mantle. Some of these investigations have been carried out using seismic information from arrays of seismometers that Professor Wysession has deployed across America. The results show that our planet is in constant internal motion, carrying heat from the deep interior up to the surface like a continual conveyor belt. Professor Wysession is also a leader in geoscience education. He is the lead author of Prentice Hall's ninth-grade physical science book, *Physical Science: Concepts in Action*. He also has supervised, in the role of primary writer, several other secondary-education textbooks, such as Prentice Hall's ninth-grade text *Earth Science* and sixth-grade texts *Earth's Interior, Earth's Changing Surface*, and *Earth's Waters*. Professor Wysession regularly gives workshops that help train secondary-education science teachers to teach earth and physical science.

At a more advanced level, Professor Wysession is the coauthor of *An Introduction to Seismology, Earthquakes, and Earth Structure*, a leading graduate-level textbook used in geophysics classes around the world. He also constructed the first computer-generated animation of how seismic waves propagate within the earth from an earthquake, creating a 20-minute film that is used in many high school and college classrooms. Professor Wysession has also written about the deep Earth in several general-audience publications, such as *Scientific American, American Scientist*, and *Earth Magazine*.

Professor Wysession's commitment to science and education began early. After he received his bachelor's in Geophysics, he taught high school math and science at Staten Island Academy in New York before going on to graduate school. After receiving his Ph.D., he joined the faculty at Washington University in St. Louis, where he has played a major role in the revisions of both the undergraduate and graduate-level geoscience curricula. He was asked to be the first Residential Faculty Fellow in Washington University's new residential college system, through which he lived with his family in a freshman dormitory for three years.

Professor Wysession has served as the editor of several journals of the American Geophysical Union, and his community service work has included several positions of responsibility within the Incorporated Research Institutions for Seismology (IRIS), which works to ensure strong continued funding for geophysical science at the national level. Professor Wysession is chair of IRIS's Education and Outreach program, overseeing the improvement of geophysical education on a variety of levels.

Professor Wysession's research and educational efforts have been recognized through several fellowships and awards. He received a Science and Engineering Fellowship from the David and Lucille Packard Foundation and a National Science Foundation Presidential Faculty Fellowship, awarded by President Clinton; both were awarded to only 20 American scientists across all disciplines. Professor Wysession also was awarded fellowships from the Kemper and Lily Foundations to enhance his teaching. He has received the Innovation Award of the St. Louis Science Academy and the Distinguished Faculty Award of Washington University. In 2005, Professor Wysession had a Distinguished Lectureship with IRIS and the Seismological Society of America, entertaining and educating audiences across the country about earthquakes and seismology.

Table of Contents
How the Earth Works
Part IV

How the Earth Works

Scope:

Because the daily lives of most people nowadays can be so busy and hectic, it is appealing to think that at least the ground beneath our feet is steady, constant, and unchanging. Nothing could be further from the truth. We live on a vibrant, dynamic planet that is constantly in motion, inside and out. If you could view Earth's history sped up, like a movie on fast-forward, our planet would look more like the swirling eddies of a whirlpool than a ball of rock. Continents would whiz about the surface, and rocks would continuously be cycling from the surface to the deep interior and back again. Because the surface changes so much over time, you would no more be likely to recognize the planet of our past than you would the planet of our future. Recent discoveries in the earth sciences (geology, geophysics, geochemistry, and geobiology) are now revealing what our planet Earth is made of, what its history has been, and, more importantly, "how it works."

The movie analogy is really not a bad one. Our current scientific investigations give us a "snapshot" of our planet as it is today. From this single image, we attempt to reconstruct its past and predict its future. It is a difficult task, like trying to reconstruct the plot of a movie like Humphrey Bogart's *The Big Sleep* from just one still. The detective movie's plot, with all of its twists and turns, is hard enough to follow with repeated viewings, but to jump in the middle and figure things out would be daunting, if not impossible; this, however, is what geologists do. They are like detectives themselves, examining the geological clues at hand in order to not only reconstruct Earth's history but also to make predictions about its future.

While it is true that our world is in flux and we may be, as Etta James sang, "Standin' on Shaky Ground," there really are some constants in our world. As far as we can tell, there are definite laws to the universe. The fundamental forces that control the motions of objects and the flow of energy seem constant and unchanging. In fact, given these laws, once the Big Bang occurred, 13.7 billion years ago, the eventual formation of stars and planets was inevitable. The machinery of our universe was set in motion; gravity, electromagnetism, and the strong and weak nuclear forces made sure that there were lots of planets orbiting lots of stars in lots of galaxies. We have a particular interest, however, in one specific planet: Earth. Though there are likely to be many billions of planets in just our galaxy alone, it turns out that very few might be like our own. The conditions required to maintain liquid water on a planet's surface for 4 billion years (the time needed for single-celled life to evolve into something that can dribble a basketball or write a love sonnet) are remarkably unusual, and I will explore this idea in more detail later on in the course.

One very important part of the study of how the earth works is the interdisciplinary nature of it. Earth science is not for the faint of heart—this is not "rocks for jocks." In a modern-day university earth science department lecture, you are as likely to hear about the biological DNA of rock-chewing bacteria, the physics of the magnetic field of Jupiter, or the chemistry of ozone reactions in the atmosphere as you are likely to hear about more traditional topics of "geology." This is because the divisions between the different sciences are entirely artificial. Nature does not know about biology, physics, and chemistry; there is only Nature, and all of the sciences are involved in it. This is nowhere more true than in the study of a planet and how it works.

In very general terms, however, Earth's story is a simple one. Earth was intensely hot when it first formed and has cooled ever since. In fact, by about 50 million years after the origin of the solar system (which we now think was about 4.567 billion years ago), Earth may have been entirely molten. Since that time, Earth has steadily cooled down, losing its heat into space. This is what all planets do, and the particular size, location, and composition of Earth (including, very importantly, the amount of heat internally generated through radioactivity) has determined *how* Earth has cooled down. For our planet, the flow of heat from the interior to the surface takes the form of plate tectonics, which involves the vigorous convection of Earth's rocky mantle layer and the horizontal motion of broken pieces (plates) of Earth's outermost layer. As the plates move, they drag the continents about the surface, and the history of these continental collisions has been largely responsible for the geology we find about us. Even today, dramatic occurrences like earthquakes, volcanoes, the opening of oceans, and the upward thrust of mountains result from the inexorable motions of plate tectonics, releasing unfathomable amounts of energy.

Any good story has to have conflict, however, and it turns out that plate tectonics has a nemesis: the sun. As fast as mountains go up and lands are formed, sun-driven erosion tears them down. Sunlight drives the cyclic flow of water through the oceans and atmosphere, and the scouring of water and ice destroys rock and carries it to the oceans. Rivers are the highways of this destruction, carrying hundreds of millions of tons of former mountains toward the oceans each year. The surfaces of the continents are therefore like a battleground torn and ravaged by the two armies of Earth's interior and the sun, each relentlessly expending their arsenals of energy upon it. At various times in Earth's history one or the other may appear to be the victor, but it is the struggle between the two, unassumingly known as the rock cycle, that has shaped the lands we live on.

There is one more frequent characteristic of a great movie: a surprise twist of the plot in the end. *We*—humans—are that surprise. It is not possible for us to examine Earth objectively as if we were something *other* than the planet we live on. We are an integral part of Earth, constantly sharing our atoms with it (there are atoms in your body that were in dinosaurs, volcanoes, Julius Caesar, and that have flowed out the mouth of the Nile River many, many times). In fact, we might be considered as Earth's experiment in consciousness. Life has always played an important role in shaping Earth's surface—on the land, in the oceans, and in the atmosphere—but we have now reached a critical moment where humans have become the dominant agent of geologic change on Earth. We are altering Earth's land, water, and air faster than any other geologic process. It is therefore vitally important that we understand, in the context of *How the Earth Works*, the nature of our geologic powers if we are to have any hope of being able to control them.

Lecture Thirty-Seven
Glaciers—The Power of Ice

Scope:

Glaciers occur where it is cold enough for snow to remain frozen and turn to ice when it is compacted by more snow. Alpine glaciers are found in mountains around the world. Continental glaciers contain most of the world's ice. Currently, Antarctica and Greenland are the only two continental glaciers. Glaciers are slowly moving rivers of flowing ice. They move through both ductile flow and sliding along their base. Snow is added to glaciers at the top, where temperatures are colder, and leave at the bottom through melting, sublimation, and calving. In between, the flow of ice is a powerful scouring agent that carves out deep valleys from the sides of mountains and deposits the pulverized rock at the bottom of the glacier into heaps called moraines. Alpine glaciers are the primary agent responsible for the rapid erosion of mountains. When continental glaciers recede, many new features of the land are formed, both erosional and depositional. For example, the "ten thousand" lakes of Minnesota are a result of recent continental glaciation.

Outline

I. Glaciers form the last of the water-based environments where rock is eroded, transported, and deposited to make new sedimentary rock.

   A. All these processes are occurring simultaneously.

   B. Glaciers take this course into a new direction: climate, because cold weather is needed to make ice.

   C. Glaciers can be remarkably efficient agents of erosion, tearing away mountains faster than any other geologic process discussed so far.

II. There are two types of glaciers: continental and alpine.

   A. Continental glaciers currently consist of Antarctica and Greenland.

   B. Alpine glaciers are responsible for reshaping land surfaces all around the globe, not just near the poles.

C. Almost all glacial ice is contained within Antarctica, a smaller amount in Greenland, and less in alpine glaciers.

III. Glaciers are continuously moving rivers of frozen water.

    A. Glaciers accumulate ice upstream in what we call a "zone of accumulation," and lose ice downstream in a "zone of ablation" or "zone of loss." The division between the two is called the "firn line."

    B. The accumulation of ice occurs through the falling of snow, which compacts and eventually turns to ice.

    C. The loss, or ablation, of ice can happen in several ways.

        1. Most of the ice is lost through melting, which happens at the bottom and underneath the glacier and helps to lubricate it and make it flow more easily.

        2. Ice can also be lost through sublimation, where water goes directly from a solid to a gas.

        3. If the glacier enters any sort of water environment, large chunks of the ice break off and float away to form icebergs. This is called "calving."

    D. Glaciers move in two different ways depending upon the slope, size, and altitude of the glacier.

        1. Glaciers move by ductile flow—actually flowing downhill. Just like a stream, the flow of a glacier is fastest within the middle and slower along its sides.

        2. Glaciers also move by basal sliding. As they slide, they tear rock off and actually shape the surface of the land.

    E. Glaciers move slowly, usually centimeters to meters per day. However, they can sometimes surge faster, as when they are reaching the ocean and flowing into it.

        1. The speed of a glacier changes with its slope.

        2. A glacier will speed up when it goes over a ledge and will crack and open up to form crevasses. If you see crevasses in a glacier, it's a sign that the glacier is increasing its speed.

    F. The style of flow for glaciers varies greatly between continental and alpine glaciers.

        1. Alpine glaciers are dominated by the presence of very narrow valleys, structures that are reminiscent of the fractal structures of streams.

2. Continental glaciers flow out in all directions, and tend to form narrow channels only when the flow goes through mountain ranges.

G. Glaciers are always flowing downhill; ice is always moving, despite its sometimes stationary appearance.

IV. As the ice flows, it erodes Earth's surface.

A. Erosion by ice plucking occurs when the ice gets into the cracks in rocks as it moves and pries rocks off, making them part of the base of the glacier.

B. Once the rock gets embedded within the glacier, it becomes an additional agent of erosion. Rocks embedded within the glacier, dragging across the surface of the ground, tear off more and more rock.

C. When ice erodes, U-shaped valleys are created, in contrast to the V-shaped valleys created by streams.

1. With a stream, most of the erosion is occurring at the bottom of the valley. As the stream cuts down, it destabilizes the slope. Rock falls down and debris flows into the stream, which, over time, widens its V-shaped valley.

2. With a glacier, the ice is in contact with the valley at all locations and so grinds out the valley into a classic U-shape, with a headwall at the top that's part of a bowl-shaped structure called a "cirque."

D. At the bottom of a glacial valley floor, you'll often find large boulders. When the boulders clog up a stream, a "glacial tarn" is formed.

E. As glacial valleys work their way up a mountain, they sometimes leave only a small peak of rock at the top, called a "horn," which has multiple cirques surrounding it. The Matterhorn in Switzerland is a classic example.

F. Above the firn line, snow continues to accumulate and the glacier is white. Below the firn line, the glacier begins to melt, rock is exposed, and the glacier appears darker.

G. Continental glaciers remove vast volumes of rock from continents through the same mechanisms as alpine glaciers, but on a much larger scale. Huge portions of the United

States are covered with thick blankets of Canadian rock scraped off during the last Ice Ages.

H. When mountains and hills aren't totally removed by continental glaciers, they are left as low, elongated, whale-back-shaped hills called "drumlins," which end up pointing in the direction the ice flowed.

I. Erosion from continental glaciers usually occurs gradually, but can be catastrophic, especially when an Ice Age is ending.

    1. Geologists found evidence of water erosion from enormous volumes of water in the northern plains region that puzzled them.

    2. It turns out that this erosion occurred from the sudden bursting of glacial lakes that formed as the climate warmed. The water traveled hundreds of kilometers in a matter of days or weeks.

V. Rock torn off lands and carried by water gets transported and deposited by glaciers, water and wind.

A. Alpine glaciers tear rock away from mountains and dump it at the end of the glacier where it forms a "terminal moraine." Glacial sediment is always completely unsorted. Giant boulders, gravel, dirt, and tiny powdered rock are all jumbled together and dumped in one place.

B. If the climate is changing and the end of the glacier is moving, the sediment can be spread out over large distances.

C. Often a huge amount of sediment dumped at the base of a glacier overwhelms the carrying capacity of the stream when the ice melts. This forms a "braided stream" in which the water winds back and forth across a huge expanse of sediment.

D. Continental ice sheets are so huge that the deposition of sediment occurs on a huge scale and the final, terminal moraines can be enormous. A good example is the line of terminal moraines off the eastern coast of North America that include Long Island, Cape Cod, and Martha's Vineyard.

E. When glaciers reach water, they often create enormous blocks of ice that float away. The blocks carry rock with them that gets deposited across the ocean seafloor. We can go back in time and find episodes of glaciation by looking at the locations of sediments taken off of the centers of continents and dropped as large blocks carried by icebergs.

VI. There are other features of deposition of sediment torn off from glaciers.

   A. One is the formation of thousands of lakes across continents. As ice breaks up on the edge of a continental glacier, it will fall in front of the glacier in the form of large blocks. As the ice sheet recedes, sediment gets dumped around them to form new land. When the ice blocks melt, they form depressions called "kettles."

   B. Another result is the creation of pulverized rock, called "rock flour," which is rock that gets ground up into very fine dust. Rock flour provides the base for very fertile soils.

Recommended Reading:

Bowen, *Thin Ice*.

Post and Lachapelle, *Glacier Ice*.

Questions to Consider:

1. Long Island and Cape Cod are in the ocean, yet sediment deposited in the ocean is usually carried away by ocean currents. How is this reconciled?

2. When two arms of a glacier come together to make a larger ice flow, there is often a black line of rock (a medial moraine) within the glacier that separates the different ice sources. How would this line form?

Lecture Thirty-Seven—Transcript
Glaciers—The Power of Ice

Welcome. In this lecture, I want to talk about glaciers. Now, glaciers form the last of the water-based environments that I'm going to talk about, where rock is eroded, transported, and deposited to make new sedimentary rock.

Now, I've talked about these different environments separately—streams, and groundwater, and shorelines, etc., but it's really only for convenience in the format of these 30 minute lectures. It's important to remember that all these processes are actually always occurring together, simultaneously. They're constantly interconnected. Glaciers, however, are going to take this course off into a new direction, and that's climate, because you need cold weather to make a whole lot of ice.

Now, I once hiked around Switzerland, and I trekked up to Grindelwald, which is a small village near the base of Mount Eiger. And you know, I still have a hard time comprehending it. I mean, I have lots of pictures that I took, but they just don't do it justice. It was, and still is, hard for me to comprehend standing in front of a vertical, 2 mile high wall of rock. It's just hard for me to comprehend how tall that really was. What happened was that several episodes of glaciation had come in, and had carved out that whole valley, and had cut straight down into the rock, forming this towering vertical wall.

Now, you don't have to go to Switzerland to see this if you've ever visited Yosemite National Park and wondered at the vertical granite cliff of El Capitan, three-fifths of a mile high. Or marveled at the shape of Half Dome, which looks like a giant knife came and neatly carved away half of it. And in a sense, that's exactly what happened, only the knife was made of ice.

Now, glaciers can be remarkably efficient agents of erosion, tearing away mountains faster than any other geologic process I've talked about so far. Removed rock has to go someplace, of course, and glaciers also provide an equally important mechanism for the transportation and deposition of sediment.

Now, glaciers occur in two general categories: continental glaciation and alpine glaciation. To get ice, you need cold, and to get cold, you can do one of two things. You can go north in the northern hemisphere, or south, towards the South Pole, in the southern hemisphere, or you can go up. You can go up into mountainous regions.

Continental glaciers, which currently consist almost exclusively of Antarctica and Greenland, contain most of the world's ice. But alpine glaciers, in mountains, are responsible for reshaping land surfaces all around the globe, and not just near the poles.

If you look at the balance of ice, almost all glacial ice is contained within Antarctica, a smaller amount in Greenland, and an even smaller amount in alpine glaciers. For instance, if the Antarctic ice sheet were to melt, sea level would rise up about 61 meters. If Greenland were to melt, sea level would go up an additional 7 meters. And if all the world's mountain alpine glaciers were to melt, the sea level would go up less than an additional meter. That's how much of the ice is stored within Antarctica.

Now, glaciers are continuously moving rivers of frozen water, and using the term "river" is really not a bad description because the ice of a glacier actually flows downhill in a fluid manner.

Now, glaciers have two different regions: They accumulate ice upstream in what we call a "zone of accumulation," and they lose ice downstream in a "zone of ablation," or "zone of loss." The division between these is identified and called the "firn line."

The accumulation of ice in a glacier occurs, really, only through one way, through falling snow. And, as you know from making a snowball, snow is initially quite loose. In fact, it can be 70% air, but as more snow gets piled on top of the upper part of the glacier, the snow begins to compact, and that air gets squeezed out, and gradually, over time, the snow turns into ice. That ice will begin to flow down, and as it flows down, it begins to warm, simply because temperatures are warmer at lower altitudes. Or, in the case of large ice sheets moving hundreds or thousands of kilometers, temperatures increase at lower latitudes.

The loss, the ablation of ice from a glacier, however, can happen in several different ways. Most of the ice is lost through melting. This happens at the end of a glacier, at the bottom, and it also happens along the base of a glacier, right underneath the glacier. And incidentally, that actually helps lubricate the glacier and helps it flow more easily at times.

Ice can also be lost through a process called "sublimation," and here, the water goes directly from a solid phase to a gas phase, skipping the phase of liquid ice. In other words, the ice turns directly into water vapor. This happens from sunlight hitting the surface of the ice, giving those water molecules, bonded tightly together as ice, enough energy to jump right into water vapor.

And very importantly, if the glacier enters into any sort of water environment (a lake, the ocean, an inlet or bay), then you get a process called "calving," where large chunks of the ice break off and float away to form icebergs. In the case of Antarctica, some of these sheets of ice breaking off can be as large as states in the United States. They can be enormous, vast areas, huge sheets that will float and drift away from the continent.

Glaciers move in two different ways, and it depends upon the slope, the size of the glacier, and also the altitude, the elevation, of the glacier. Glaciers move by ductile flow, as I've already talked about, actually ice flowing downhill. And we shouldn't have a problem with this now, because we saw this with Earth's mantle in convection. Right? Rock flows (over geologic time).

Well, ice is solid, but it also flows, and in fact, it flows a lot more easily than rock does. You can actually see this over time, and geologists sometimes will put long poles down into the middle of a glacier and watch over time as that pole bends and warps as the glacier flows. Just like a stream, it turns out that the flow of a glacier is fastest along the middle and slower along its sides.

But—and this turns out to be really important for erosion—glaciers also move by basal sliding. And, as I mentioned, this is helped by the presence of water that often is found along the base of the glacier. So, in other words, the ice isn't just flowing along the bottom; it's actually sliding across the bottom. And as it slides along, it tears the rock off the bottom, and that's what makes glacier so important for shaping the whole surface of the land.

Now, glaciers move slowly by our standards, usually centimeters to meters a day. If you're standing by most glaciers, you don't see them move. But interestingly, you can actually hear them, because as the ice is flowing and moving, it cracks tremendously, and you often hear very loud creaks and groans as that ice flows. However, there are cases where glaciers can surge 100 times faster than these rates, especially when they're reaching the ocean and flowing into the ocean. And in this case, you can very definitely see the ice flow.

Now, just like with a stream, the speed of a glacier changes with its slope. The flow of the glacier will speed up when it goes down over a ledge. In this case, the ice, which isn't as fluid as water, will crack and open up to form crevasses. If you see crevasses in a glacier, it's a sign that that glacier is increasing its speed. Usually it's going downhill over some ridge or ledge. As the glacier slows again, those crevasses will close up.

Now, the style of flow for glaciers varies greatly between continental and alpine glaciers. Alpine glaciers are dominated by the presence of very narrow valleys that are carved out by streams of ice. And in fact, if you look at an aerial view of an alpine glacier, you see a structure that's sort of reminiscent of the tree-like fractal structures of streams. You often have many smaller ice streams up in all the different valleys that will often converge and come together to form one large, major ice stream.

With continental glaciers on Antarctica and Greenland, flow, however, generally goes out in all directions from the regions of highest elevation, and they tend to form narrow channels only when the flow goes through mountain ranges. However, this is not as unusual as you would think, because think about what's happening in a place like Greenland or Antarctica. You're taking the crust, and you're stacking kilometers of ice on top, largely in the middle, less at the edges. The result is that the middle of Greenland and Antarctica are actually pushed down. In fact, if you could suddenly, instantly remove all the ice, you'd find that the surface of these continents, in many places, is below sea level.

Remember I talked about how Canada is rising up; it is actually lifting up at about a millimeter a year. Well, you put all the ice on top of Antarctica, it pushes the continent down, pushes the asthenosphere out of the way. That means that the edges of the continents tend to be at a higher elevation, and there often are mountain ranges at the edge of Antarctica and Greenland, and sometimes the ice has to flow out of the channels in between these mountains.

Now, there's a common misconception about glaciers. It turns out that a glacier is always flowing downhill. Ice is always moving. Sometimes, people look at a glacier where the front may be stationary over time. Or if climates are warming, the front actually may be receding, going uphill, in which case, you're losing ice faster than you're adding it. But the ice is always continuously flowing downhill.

As the ice flows, it erodes the surface, and it does this in two primary ways: through a process called "ice plucking" and through abrasion. It's kind of similar to what happens on the bottom of a stream. Ice plucking involves the ice moving, and actually getting into the cracks in the rock, and prying rock off, and lifting it up, and making it become part of the base of the glacier. It's sort of like water flowing along the bottom of a stream and, by hydraulic action, lifting up cobbles.

However, once that rock gets embedded within the glacier, it becomes a whole additional agent of erosion. Now you've got rock embedded in the glacier, dragging across the surface of the ground underneath, tearing off more and more rock, accelerating the rate of erosion. It's sort of like putting diamond chips into a drill bit to make it be able to drill faster and harder.

At high elevations, this process of erosion occurs year round. There's ice found all year round at places like Mount Kilimanjaro in Africa and in the Andes in South America. These are right at the equator, and yet they have ice year round, because they are high enough in elevation.

And when the ice erodes, they create very distinctive glacial valleys. Actually, you can always tell when a valley has been shaped by a glacier, as opposed to a stream, because the glaciers make U-shaped valleys, curved out, as opposed to streams, which make V-shaped valleys.

Why does this happen? Well, with a stream, all the erosion, or most of the erosion, the removal of rock, is occurring in the stream at the bottom. As the stream cuts down, it destabilizes the slope, which makes the angle of the land on either side exceed its angle of repose. You get slumping and landslides, rock falls, debris flows, carrying into the stream then out, and the stream over time widens its V-shaped valley.

With a glacier, you have ice in contact with the valley at all locations, and it just grinds out that valley in this classic U-shape. At the top of one of these glacial valleys, you'll often find a headwall that's part of a bowl shaped structure. We call that a "cirque." In fact, if you've gone skiing in places like the western U.S., you often are skiing in these cirques, these bowls at the top of the glacial valleys. At the bottom of the valley floor, you'll often find boulders that are larger than houses and that have fallen off the side of the valley. And sometimes, these boulders actually clog up a stream, and you'll get a lake called a "glacial tarn."

As these glacial valleys work their way up a mountain on all sides, sometimes all that's left of the mountain at the end is a small peak of rock with cirques on all sides. We call that structure a "horn," and the classic example of that is the Matterhorn in Switzerland.

Alpine glaciers contain large amounts of rock within them that have been torn off, and as they carry that rock through the glacier, downward, you can actually see it. In fact, in many cases, if you look at ice streams from alpine glaciers, you can see black lines extending throughout the glaciers. And if you notice, all these black lines head back to a place where two ice streams come together. What happens is that, as they're eroding, they're tearing rock off the side walls, and when you bring two streams together, you now have a line of that eroded rock embedded within the glacier itself.

Also, you can often identify the location of the firn line, because above the firn line, where snow is still accumulating, the glacier is white. Below the firn line, where you're beginning to lose that glacier through melting and sublimation, the rock that has fallen down on top of the glacier now begins to be exposed, and the glacier often will begin to get darker the further down you go in elevation.

These flowing ice streams can sometimes cut each other off, and the result, once the ice is gone, are these hanging valleys where one valley comes out and drops suddenly into another glacier valley. These sometimes can make very dramatic waterfalls. I previously mentioned that the classic example is found in Yosemite—Bridal [Veil] Falls.

Over time, if these glacial valleys are at sea level, and they get flooded (because, as I previously mentioned, the global sea level has gone up nearly 400 feet since the end of the last Ice Age), these become fjords. So, throughout Scandinavia, the fjords that extend up, these arms of the ocean that extend up into the land, are glacial valleys that formed through ice streams flowing out, and they're now flooded by the ocean.

Continental glaciers remove vast volumes of rock from a continent. In fact, it turns out that the U.S. is made of largely foreign soil. Huge portions of our country are actually covered with thick blankets of Canadian rock scraped off during the last Ice Ages. The [continental glaciers] just tore these Canadian rocks and mountains away, carried them down across the border and dumped them in North America.

Continental glaciers remove rock and soil through the same mechanisms as alpine glaciers, through this process of abrasion and ice plucking, only it does it on an enormous scale. Where mountains and hills aren't totally removed by these continental glaciers, they often are left as small stretched out hills, low, elongated sort of whale-back shaped hills that we call "drumlins" that end up pointing in the direction that the ice flowed.

You go across New York State or the Dakotas, and you find large numbers of these drumlins pointing north-south—remnants from the last Ice Age. Erosion from continental glaciers usually happens in a gradual manner but can sometimes be catastrophic, especially when an Ice Age is ending.

Let me explain. When geologists visited the northern plains of the United States, they found something very puzzling. Now, stream beds—I've talked about the process of streams—have a very distinctive appearance on the bottom of a stream. You often get ripples forming from the process of saltation, just like with a sand dune. You've often walked, probably, across the bottom of a pond or a stream and felt these ripples on the bottom, and stream ripples are usually centimeters to meters apart, depending upon the speed and volume of a stream.

Well, the northern plains region showed evidence of former water erosion, only the ripples were ridges that were 100 meters apart. These ripples had football fields worth of length in between them, and it's hard to imagine where such enormous volumes of water could have come from, but it turns out that these occurred from the sudden bursting of enormous glacial lakes that formed as the climate warmed and the ice began to melt, but was dammed up by ice.

Occasionally these dams would break, and lakes the size of the Great Lakes would rush out and drain across the surface, going hundreds of kilometers in a matter of days or weeks. One of the best known regions is found in eastern Washington State and is called the Channeled Scablands, and this shows evidence of an unbelievably huge amount of water rushing across this region, just tearing the surface away—again, forming ripples 100 meters apart.

I once had a first-hand experience of this when I visited the St. Croix River, which forms the boundary between Minnesota and Wisconsin. I went canoeing there, and now there's such a small amount of water trickling down the stream that, frankly, you can canoe just as easily upstream as you can downstream. However, if you look around there, you see evidence of an enormous amount of water in the past—very steep huge cliffs on either side.

Near the stream, there are huge potholes that are sometimes 40 feet deep, sudden holes in the ground, perfectly round. And often, at the bottom of these potholes, you'll see a boulder the size of a small bus that's perfectly rounded. These potholes formed when the glacial lake Duluth drained in about a week. It was one of these dammed-up glacial lakes at the end of the last Ice Age, and the water burst out, it carved the steep cliffs on either side, and in the process, the water flowed so quickly that wherever there was an eddy, a giant boulder would drill down, making potholes that were sometimes 40 feet deep.

Now, all this rock torn off lands, carried by water, has to go someplace, and it gets transported and deposited not only by the glaciers, but also by water and wind. Now, in the case of an alpine glacier, most of the rock that gets torn off the mountains gets dumped right at the end of the glacier, and that's called a "terminal moraine." You can always identify glacial sediment, because it's totally unsorted. You've got giant boulders, and gravel, and dirt, and tiny powdered rock all jumbled together and dumped in one place.

If the glacier stays put for a long time—in other words, if the end of the glacier doesn't change—you build up a huge terminal moraine there. But if the climate is changing, and the end of the glacier is moving its location, then the sediment can be spread out over fairly large distances.

There's often a huge amount of sediment dumped right at the base of a glacier, and it overwhelms the carrying capacity of the streams. In other words, the ice is melting to form water, but there's too much sediment, and the streams are totally sediment-choked for great distances away from the glacier.

You can often identify this from a particular type of stream called a "braided stream." If you ever look at these from an aerial photograph or from an airplane, you can see the water winds back and forth across a huge expanse of sediment.

Now, because continental ice sheets are so huge, the deposition of sediment occurs on a huge, grand scale, and the final moraines, the terminal moraines where the sediment gets dumped, can be enormous. They can extend for hundreds to thousands of kilometers.

In fact, Long Island, New York, Cape Cod, Massachusetts, and all the islands in between—Martha's Vineyard, Block Island, all of these—are all part of the same single terminal moraine from an ice sheet that ended about 21,000 years ago. So, in other words, Long Island, that whole island, is nothing but glacial debris dumped there at the front of the glacier.

Now, it's off the coast of the land, now. Why is that the case? Well, remember, back when it formed, that wasn't the edge of the continent. Sea levels were much lower, and that was on land. And so, that represented the intersection of the glacier with the sea, and it extends all the way from New York past Boston. It's all the same moraine, though Red Sox and Yankee fans might be reluctant to admit it.

Now, when the glaciers do reach the water, they often create enormous blocks of ice that float away, icebergs, and these obviously have been significant hazards for ships long before the Titanic sank. Interestingly, from a geological perspective, these icebergs still carry a lot of rock within them, and that rock, as a result, can float large distances across the ocean and can get deposited across the ocean seafloor out in the middle of the deep ocean. And it's very interesting, because if you find Canadian rock out in the middle of the Atlantic in a layer there, you know that must have occurred during a period of glaciation.

And we can actually go back in time and find episodes of glaciation by looking at the locations of sediments, sediments taken off of centers of continents and dropped, not as mud, the way normal sediment forms in the middle of the ocean, but as large blocks carried from icebergs. I'll talk later on in the course about how there was a period in our history, about 800 to 600 million years ago, where the earth almost froze over. In fact, it may have become entirely frozen. These glacial sediments are found scattered throughout the oceans.

Now, there are other interesting features of the whole process of deposition of sediment torn off from glaciers. One interesting result is the formation of thousands and thousands of lakes across continents. It's an interesting process. If you go to the edge of a continental glacier, as the ice is breaking up, it will begin to fall in front of the glacier, and the ice sheet begins to recede, leaving these large blocks of ice that are slowly melting. Dumped around them is all the sediment of the outwash plain. Well, the sediment settles. It forms land. The ice melts and forms a depression, we call that a "kettle," that forms once the ice is gone and the sediment deposits all around it.

Minnesota is known as the "Land of Ten Thousand Lakes," and if you look at a map of Minnesota, that's not an exaggeration. The whole surface of the state is covered with lots of tiny, little lakes. These all represent, or most of them, pieces of ice that had glacial sediment deposited around them. Of course, as I mentioned previously, sedimentation will fill in lakes, so over time, these lakes will get filled in. In fact, many of them already have.

There's another interesting result of the process of glaciation. It's the creation of pulverized rock, what we call "rock flour," rock that's entirely ground up into very fine dust by the continued motion of the glacier. Well, when the glacier leaves and the water dries up, this rock flour can get blown hundreds to thousands of kilometers as dust, and much of the Midwestern United States is covered by thick layers of this. It's called "loess," and across Indiana, Illinois, Ohio, it makes for tremendously fertile soils.

Now, you might never have realized that the thick soils of Illinois, the thousands of lakes of Minnesota, and the very existence of Long Island and Cape Cod as glacial terminal moraines are all the result of the ice from the last Ice Age. In fact, in some parts of the country, Route 70, stretching across the whole country, represents the southernmost extent of the most recent glaciers. Chicago, at that time, was buried under more than a kilometer of ice.

But we're left with a very obvious question: Why did it end? Where did all this ice go? What brought about the end of the Ice Age? Why is it that Greenland is the only remaining part of massive ice sheets that, until very recently, covered huge portions of North America and Eurasia?

Well, in the next lecture, I'm going to begin to unravel this very complicated story of climate change, and I'm going to talk about the huge effect that it has had. Not only on the surface of the land, but on the very course of human history itself.

Lecture Thirty-Eight
Planetary Wobbles and the Last Ice Age

Scope:

Strange as it sounds, the shape of Earth's orbit affects the size of mammals. There is currently a noticeable cyclical nature to the alternation of cold glacial periods and warmer "interglacials," and this is predominantly due to variations in Earth's orbital characteristics (called "Milankovitch cycles"). Changes in the ellipticity of Earth's orbit around the sun, as well as changes in the tilt and direction of Earth's axis of rotation, affect the amount and distribution of sunlight that Earth receives. This affects global climates and, when it is cold, large ice sheets cover many parts of the continents. The changing amount of ice has some very serious consequences. During the last Ice Age, the sea level was more than 120 meters (400 feet) lower than it was today, increasing the area of land and allowing humans and other mammals to migrate between continents. The ice sheets destroyed most high-latitude environments, and favored the evolution of larger mammals (which lose heat more slowly with lower surface-area-to-volume ratios) like giant mammoths.

Outline

I. The important questions for climate are: What are the systems involved and over what scales do they operate?

    A. Some time scales of temperature change we understand fairly well.

        1. The daily cycle of temperature: It's warmer during the day than at night because there is more sunlight.

        2. It's warmer during the summer than in winter.

    B. Some time scales are more complicated.

        1. Factors that affect global temperatures on scales of tens to hundreds of years include the output of energy from the sun and ocean circulation patterns.

        2. Over scales of tens of thousands to hundreds of thousands of years, factors that control climate include Earth rotation around its axis and its orbit around the sun.

3. Over scales of tens of millions to hundreds of millions of years, factors that drive climate change include plate motions, large scale erosion of mountains, and long term distribution of carbon dioxide cycling between the atmosphere, ocean, and land.

C. The climate may be going up on one cycle and down on another, and humans are doing all sorts of inadvertent experiments with this system as well.

II. Why there are seasons.

A. Earth rotates on its axis in a clockwise direction (assuming that the North Pole is up). That is why the sun rises in the east.

B. The Earth takes 23 hours, 56 minutes, and 4 seconds to make one full rotation around its axis. It has moved along part of its revolution around the sun during this time, so has to rotate an additional 3 minutes and 56 seconds to bring the sun back to the same spot overhead and complete a 24-hour cycle.

C. The seasons occur because Earth's axis is tilted with respect to the plane that it travels in during its revolution around the sun.

D. The northern hemisphere has summer when it is facing the sun (winter in the southern hemisphere). Six months later the planet is on the other side of the sun and the northern hemisphere is facing away from the sun. That causes winter in the northern hemisphere and summer in the southern hemisphere.

E. The characteristics of Earth's rotation around its axis and the revolution around the sun change slightly over scales of tens of thousands to hundreds of thousands of years. This is why we have Ice Ages.

F. For the most part, life is dependent upon the existence of sunlight. Life on Earth exists in a fairly delicate balance concerning the sunlight that it gets. Small fluctuations in Earth's orbit change the amount and distribution of sunlight across the globe, and this can have serious effects on global climates over intermediate time scales.

III. When we look at a record of global temperatures over the last few hundred thousand years, several patterns become apparent.

  A. Our climate occasionally jumps up into brief stable periods of warm temperatures called "interglacials."

  B. Interglacials are followed by a gradual decline in global temperatures toward colder climates and also, more importantly, very variable climates.

IV. It is challenging to try to figure out past temperatures. We use other kinds of data, called "proxies," to provide the past record of temperature change.

  A. One important method is to drill into ice cores in Greenland and Antarctica. Ice within these two glaciers can be as old as a million years in places.

   1. This technique looks at different isotopes of hydrogen atoms and tracks the presence of a rare hydrogen isotope called deuterium.

   2. Elevated levels of deuterium suggest that snow fell during a warmer climate when the temperatures of the air and water were higher.

  B. We can also get temperature change information from a technique using marine shells to analyze ratios of oxygen isotopes in the oceans.

  C. During Ice Ages, the ocean was rich in O_{18}, so fossil shells from that time are also rich in O_{18}.

  D. The deuterium isotope values in ice cores and the O_{18} isotopes in marine fossil shells give us very similar results. This gives us a sense that we are accurately measuring ice volume and temperature going back over hundreds of thousands of years.

V. The pattern of climate change is related to the slight changes in Earth's orbital parameters (called "Milankovitch cycles") which occur in three ways and affect the amount and distribution of sunlight that the earth gets.

  A. The ellipticity (eccentricity), obliquity (tilt), and axis of Earth's rotation all change.

B. During a planet's elliptical orbit around the sun it moves fastest when it is closest to the sun and slower when it is farther away.

C. The more elliptical a planet's orbit, the less total sunlight a planet will receive over the course of a year, and the colder it will be.

D. The range of the oscillation is from about 95,000 years to about 400,000 years, with a dominant period of a little more than 100,000 years. We saw in the records of temperature change that 100,000 years is roughly the amount of time between interglacial periods.

E. Earth's axis is currently tilted 23.5 degrees with respect to the ecliptic. The greater the axis is tilted, the greater the temperature difference between winter and summer.

F. The Earth's North Pole currently points at a star we call Polaris, the North Star. In another 12,000 years, the axis will point in a very different direction toward the star of Vega. This is called axis precession.

　　1. The precession does not change either the amount or the distribution of sunlight.

　　2. However, it's timing in relation to the change in ellipticity of Earth's orbit and the tilt of the axis does effect climate change.

　　3. Whether the North Pole is pointing away from or toward the sun when Earth is farthest away from the sun changes the way sunlight gets distributed around the planet.

G. All three factors play a role, but it is most significant that the times of the interglacials line up with the pattern of changes in the ellipticity of Earth's orbit.

H. These small variations in Earth's orbital parameters are what cause Ice Ages. The most recent Ice Age was about 70,000 to about 10,000 years ago and, at times, covered much of North American and Eurasia with ice. It also caused much of the dramatic geology discussed in this course.

VI. Times of global cooling like these last Ice Ages have very important effects on life and living organisms.

A. As global climates cooled, regional climate bands all around the planet were shifted.

B. Alpine environments would have dropped to much lower elevations and zones of temperate forests would have extended to lower latitudes in areas that weren't entirely covered with ice.

C. During the recent Ice Ages, cold favored the evolution of larger animals because larger animals retain heat more efficiently.

D. The drop in sea levels during the Ice Ages resulted in enormous expanses of exposed land, which, in turn, resulted in increased rates of erosion.

E. The glacial sea level drop also caused fertile biologic marine environments to be quite narrow, limited to steeper continental shelves.

F. The resulting larger continents would have been dominated by monsoon-type climates, with a large annual alternation between rainy periods and dry periods.

G. When sea levels drop, continents become connected. Eurasia, North America, and South America were one continuous land mass.

    1. Twenty thousand years ago, wooly mammoths that had evolved in Eurasia crossed from Siberia into North America.

    2. Humans followed soon after into North America from Asia, about 14,000 years ago, bringing their domesticated dogs with them. In Southeast Asia, the lowering of the sea level connected Australia with the rest of Asia. Ancestors of native Australian aboriginals came over at that point, able to walk across or traverse very shallow regions.

H. This all came to a rapid end 10,000 years ago because of the effects of the Milankovitch cycles. This marked the end of the glacial period and the start of the next interglacial, and coincided (not coincidentally) with the start of human civilization.

Recommended Reading:

Broecker, *"Glaciers that Speak in Tongues and Other Tales of Global Warming," Natural History*, October 2001.

Macdougall, *Frozen Earth: The Once and Future Story of Ice Ages.*

Questions to Consider:

1. Can oxygen-18 isotope age dating work for freshwater fossil shells from lakes? Why or why not?

2. How would the composition of the ocean change after the end of the Ice Age?

Lecture Thirty-Eight—Transcript
Planetary Wobbles and the Last Ice Age

Welcome. In this lecture, I want to talk about Ice Ages, but I'm also going to start the discussion about climate change and the role it plays in shaping Earth's surface and, it turns out, in controlling the very course of human history.

Now, as an Earth scientist, I often get asked the question, Where is the climate going? Well, I can't give a simple answer to that. It just doesn't work that way. Suppose you worked with financial stock markets, and someone asked you, "Where is the market going?" How would you answer? You first might answer by asking questions back. Which market? Which sector? Are you talking short term, or long term, or intermediate term? Are you talking global markets, a regional market? Even then, it wouldn't be easy, because a particular sector might be heading in one direction, but you might know that there are other factors aligning that will cause it to change and go in another direction.

The same situation applies for climate change. I mean, when you ask about climate change, do you mean temperature? And of the temperature, do you mean annual mean temperature, the annual range of temperature? Or is it rainfall? Or vegetation? Are you talking global climate or regional climate? I mean, for instance, yes, global climates have warmed over the past century, but not everywhere. Some places have actually gotten cooler during this time. It's not a simple question.

And if you are talking global, how do you even measure that? Are you talking about land based measurements? Ocean based measurements? Measurements in cities? Measurements of temperature taken from satellites? Taken on land?

How do you measure the mean stock market? Are you talking about Dow Jones? Index 500? The Wiltshire 5000? You can see these are difficult questions.

And here are the big questions for climate: What are the systems involved and over what time scales do they operate? Because it turns out that climate lies at the intersection of so many of the different parts of Earth science that we've already talked about: the deep

Earth, the atmosphere, the ocean, the sun and the rest of the solar system, chemistry, physics, and, of course, life itself.

And as for time scales, well, climate change occurs over cycles, within cycles, within further cycles. Now, some of these we understand fairly well. It's warmer during the day than it is at night because there's more sunlight. There is a daily cycle of temperature that we understand quite well. There is also a seasonal change in temperature. It's warmer in the summer than it is in winter.

But there are also cycles that get a little bit more complicated. For instance, temperatures on a global scale seem to vary on scales of tens to hundreds of years. This involves a variety of factors that include the output of energy from the sun, or ocean circulation patterns, the water cycling back and forth between the different oceans that we already talked about.

Well, over scales of tens of thousands to hundreds of thousands of years, it turns out there are other factors that control climate, and these are totally different. This is the way that Earth spins on its axis and the way it orbits around the sun. And if you look at scales of tens of millions to hundreds of millions of years, you have a totally different set of factors that are driving climate change—things like plate motions, large scale erosion of mountains, and the whole broad, long term distribution of carbon dioxide cycling in between the atmosphere and the land and the ocean. I mean, we are just beginning to understand these properties.

So, where's the climate going? Well, it's going up on one cycle, it's going down on another, and, of course, now we humans are doing all sorts of inadvertent experiments with the system as well. It's just not an easy answer. Over the next few lectures, I hope to give you a sense of the complexity of the whole system of factors that we call climate, and I also hope to give you a sense of just how important this is for us.

Now, humans are certainly having an effect on climate and climate change in a variety of ways, and unfortunately, this has become a fairly politicized subject. I will talk about the role of humans in global climate change, but not until Lecture Forty-Five, not until I've talked about how the earth's climate changes on its own without our help.

Now, by the way, you do know why there are seasons, right? I mean, I ask because there was an infamous study of graduating Harvard students who were asked all sorts of questions, like why does the sun rise in the east and why are there seasons, and, frighteningly, large numbers of them got these questions wrong.

So, first of all, Earth rotates around an axis, and it rotates in a clockwise direction if you assume that the North Pole is up. And this is why the sun rises in the east.

Now, let me ask another question. How long does it take Earth to make one full rotation around its axis, 360 degrees? No, actually it's not 24 hours. It's 23 hours, 56 minutes and 4 seconds, because during that time the earth has not stayed still. It has moved part of its way around the sun in its revolution around the sun, and so the earth has to rotate an additional 3 minutes and 56 seconds to bring the sun back to the same spot overhead, and that gives you the full 24 hours.

Now, the seasons turn out to be a result of the path that the earth takes in its revolution around the sun. All planets move in elliptical orbits, not circles. This was the discovery of Johannes Kepler, and the earth is no exception. So, is this why are there seasons? Well, no. The reason we have winter is not because the earth is farthest from the sun at that point. In fact, it turns out that the earth is actually closest to the sun around January 4th each year. The seasons occur because the earth's axis is tilted with respect to the plane that it travels around the sun, what we call the "ecliptic." That's why all the globes are always shown with the earth tilted at this angle. This is 23.5 degrees, and so Earth's axis is not perfectly perpendicular to its path around the sun; it's slightly tilted.

So what happens is, when the North Pole points towards the sun, the northern hemisphere gets more sunlight; that's our summer. The southern hemisphere is pointing away from the sun; that's their winter. Six months later, when the earth is on the other side of its orbit, the northern hemisphere is pointing away from the sun; that's our winter. At the same time, the southern hemisphere is pointing more directly toward the sun; that's their summer.

So, now we know why there are seasons. Now, what does this all have to do with Ice Ages? Well, here's the key: The characteristics of Earth's rotation on its axis and the revolution around the sun change slightly over time. Over scales of tens of thousands to hundreds of thousands of years, and this is the reason that we have Ice Ages.

Now, our life is nearly totally dependent upon sunlight. I mean, there are a few exceptions I've talked about; there are bugs that live down in the rock, there are worms at mid-ocean ridges that feed off of food and systems that aren't photosynthesis based. They don't use sunlight, but for the most part, life is incredibly dependent upon the existence of light. And the existence of light coming to the earth is slightly variable. So, life on the earth exists in a fairly delicate balance concerning the sunlight that it gets.

Now, I don't want to alarm you. Earth's orbit is fairly steady. It's not going to fly off into space, but it does have very small fluctuations that change the amount and the distribution of sunlight across the globe. This can have very serious effects on global climates over intermediate time scales.

Now, let's look at a record of global temperature extending back a few hundred thousand years. When you do that, there are several patterns that become immediately apparent. It seems to happen that—again, if you look over the scales of hundreds of thousands of years—our climate occasionally jumps up into brief, stable periods of warm temperature. We call these "interglacials," and these are followed by a decline in global temperatures to colder climates, but also, more importantly, very variable climates. The temperature changes quite rapidly over short time scales.

So, the last interglacial occurred about 130,000 years ago, and it lasted fairly briefly. The temperature was fairly stable. The temperatures dropped, and we went into 70–80,000 years of colder variable temperatures. Then, starting a little after 20,000 years ago, the temperatures began to warm rapidly, and from about 10,000 years ago until now, we've been in another one of these warm interglacial periods—warm temperatures, but more importantly, very stable temperatures compared to the way climate change usually occurs during the periods of Ice Ages. Just before we had an interglacial, we had an extended period of Ice Ages—very cold, very variable times. So, this is the pattern over the scale of tens to hundreds of thousands of years that we need to be able to explain.

Now, it's really challenging to try to figure out past temperatures, because we weren't there with thermometers. So, we have to use other methods. We use what are called "proxies," and these are other techniques that give us the temperatures in the past. Now, one very important method is using ice cores in Greenland and Antarctica, the two large continental glaciers, because the ice has been sitting there for up to a million years in places. So, if we can drill down, we can figure out what the climate was in these different times.

This method looks at the atmospheric compositions, because it looks at the rain that fell out, or the snow that fell out, onto the glaciers, compacted into ice. It gives us long term changes in temperature going back many hundreds of thousands of years. Now, the drawback is that it's only giving us information at Greenland and Antarctica. However, those are very good measures.

It uses an unusual technique that looks at the different isotopes of hydrogen atoms within the water molecules. You see, hydrogen usually consists of 1 proton and an electron, and because the weight of electrons is almost insignificant, the weight of that is essentially just the 1 proton. However, occasionally, rarely, an isotope of hydrogen also contains a neutron. We call that deuterium. So, it has a proton and a neutron, and it's twice as heavy. Now, there aren't many of these, only 1 in every 6,500 hydrogen atoms has that neutron and forms deuterium. However, they exist in quantities enough that we can measure. And here's the interesting thing: If you've got deuterium sitting in the ocean, it takes more energy to lift a heavy atom, and therefore, more deuterium gets evaporated when the water is warm and when the air is warm.

So, if you go to Antarctica, and you drill down, and you look at the composition of the ice, where the ice contains more deuterium means that there was snow that fell on Antarctica in a warmer climate. The temperatures of the air and the water at that time were warmer.

Now, we can also get temperature change in a more global sense, and this comes from a very different technique. We use the ratios of oxygen isotopes in the oceans. You see, most oxygen occurs as O_{16}. This is an isotope of 8 protons and 8 neutrons. However, there are also small amounts of O_{18}. This consists of 8 protons and 10 neutrons. So, it's a little bit heavier. That means that when water evaporates off the surface of the ocean, very little of the O_{18} gets evaporated. So, what does come out is water, H_2O, that contains the O_{16}.

So, you have an ocean with mostly O_{16}, some O_{18}. You take the O_{16} out, you drop it on land as snow. Over time, the ratio of O_{18} to O_{16} in the ocean will get higher. There'll be proportionately more O_{18}. Now, this would be fine, except how do we actually have a record of it? We don't have ice forming from this time all around the oceans.

Well, we do have another trick, and that is that marine shells from organisms contain oxygen. Most marine shells, clams and other shells are made of calcium carbonate. This is calcium, and carbon, and 3 atoms of oxygen. So, if you look at fossils that formed at different times, and you look at the oxygen atoms, and you look at how many of them are O_{16} and how many of them are O_{18}, you get a sense of global climates. During Ice Ages, the ocean was rich in O_{18}, so fossil shells from that time are rich in O_{18}.

So, marine fossils end up being very good for global climates. Why? Because the oceans are well-mixed over very short time scales. Remember I said how the whole pattern of ocean global circulation may take less than a thousand years. Well, when we're looking at scales of hundreds of thousands to millions of years, that's a fairly short amount of time to totally mix an ocean.

Now, the deuterium isotope values in ice cores and the O_{18} isotopes in marine fossil shells give us very similar results, and that's really important, because we have totally different data in different places, and they give similar results. So, that gives us a sense of reliance that we have a fairly good measure of the ice volume and the temperature going back over hundreds of thousands of years.

If you look at this figure, the top two curves are ice core measurements of the deuterium isotopes from Antarctica. The bottom curve is the O_{18} isotopes in a sediment core of the ocean, a measure of ice volume. And you can see that, indeed, they line up very well, and we get that same pattern. Every 100,000 years or so, there's this spike of warm climate, a sudden rise in temperature. It lasts a short amount of time, and then it goes back down into an extended period of cold temperatures of essentially worsening Ice Ages until the sudden rapid rise again.

Now, I said that the pattern of climate change is very related to Earth's rotation around its axis and revolution around the sun. Well, where does this come into play? The slight changes in Earth's orbital parameters, which we call the "Milankovitch cycles," by the way—named after a Milutin Milankovitch—they occur in three ways, and taken together, they affect the amount and distribution of sunlight that the earth gets.

First of all, the ellipticity of Earth's orbit changes. That's also called the "eccentricity." And the obliquity of Earth's axis of rotation changes. Obliquity is a fancy word for tilt. And that axis of rotation also precesses. It moves around. It doesn't point in the same direction all the time. Let me break these down. Earth's orbit, as I said, is an ellipse. Now, it's nearly circular, but it's still elliptical, and more importantly, it oscillates over time.

Now, why is this important? It turns out that if you have an elliptical orbit of—let's say—a planet around the sun, as the planet moves close to the sun, it moves very quickly. But when it moves out on its elliptical orbit at the far end, it travels very slowly. When it comes back in, it speeds up again.

That means that the more elliptical your orbit is, the more your planet will spend time far away from the sun. More elliptical, less total sunlight over the course of the year—less sunlight means colder temperatures.

Now, that ellipticity actually oscillates over long periods that are much longer than an Earth year. So, for instance, the earth is traveling around the sun quickly, but that orbit is oscillating in its ellipticity. And it's a very complicated process. It's influenced by the gravitational attraction of other planets in our solar system, as well. The range of that oscillation, and the period length, is from about 95,000 years to a little bit more than about 400,000 years. But, the dominant period ends up being a little more than 100,000 years. Does that sound familiar? That's the amount of time in between these interglacial periods, just like we saw in the temperature change records.

Now, the second one is the tilt of Earth's axis. I said that Earth's axis is currently tilted 23.5 degrees with respect to the ecliptic, but that ranges from 22.1 degrees to 24.5 degrees. It doesn't sound like a lot; it's just a couple degrees. However, when your axis is more tilted, you get stronger seasons. In other words, there's a greater annual temperature change between winter and summer.

Obviously, if the axis were perfectly perpendicular to the ecliptic, there wouldn't be any seasons at all. If it were very tilted on its side, there would be dramatic changes. This tilt oscillates with a period of 40,000 years, 41,000 years and also 54,000 years, and combined together, it's a dominant period of about 41,000 years.

Now, the third part I mentioned—the precession of Earth's axis—is quite interesting. It turns out that, currently, our North Pole points at a star that we call Polaris, the North Star. It doesn't always stay that way. In fact, over about 12,000 years, the axis is actually going to point in a very different direction; it's going to point close to the star of Vega. And in another [12,000] years or so, it's going to come back again and will point toward Polaris again.

So, the axis precesses. It moves around in a circle as the earth spins out in space. And you know what? You probably have some familiarity with that if you've ever taken a top and spun it. What happens when it spins? The axis actually tilts. The path, sometimes, a top takes might be a little bit of a circle, and that axis will tilt back and forth.

Well, our planet is a spinning top in space; it does the exact same thing as this top, only it takes a whole lot longer. Now, the precession of the axis by itself, if you're listening carefully, doesn't change either the amount or the distribution of sunlight. However, its timing in relation to the change in ellipticity of Earth's orbit, and the tilt of the axis, does have effects on climate change.

In other words, is it pointing towards the sun when the earth is farthest away from the sun or is it pointing away from the sun? These factors do change the way sunlight gets distributed around the planet.

Well, if you take all of these together and combine them, what you see is a very interesting pattern of climate forcing. So, if you look at this figure, you see the top curve represents changes in climate due to changes in temperature that would be recorded at some point in the northern hemisphere due to changes in the precession of the axis, the combination of the 19, 22, and 24,000 year periods.

The next one, the green curve, is the obliquity, how tilted that axis is over a period of abut 41,000 years. The blue curve represents the change in Earth's orbit, its eccentricity, its ellipticity. And if you combine the three together, you get the yellow curve, which represents the temperature change, the solar forcing, at some location, a latitude of 65 degrees north in the summer, just chosen as an example.

When you combine them together, you see there is actually quite a strong similarity to many of the changes in climate that are represented in the bottom curve, which shows the stages of glaciation. All these different factors play a role, but notice, most significantly, that the times of the interglacials line up very well with that blue curve, with that pattern of the ellipticity of Earth's orbit.

And so, there we have it. It is these small variations in Earth's orbital parameters that end up driving temperature changes over the scale of hundreds of thousands of years and are what cause the Ice Ages.

Now, the most recent Ice Age extended from about 70,000 to about 10,000 years ago and, at times, covered much of northern North America and Eurasia with ice. It created so much of the dramatic geology that I've already talked about in this course. Large scale continental glaciation caused extreme erosion and deposition that I described in the lecture on glaciers and elsewhere.

Remember the lakes in Minnesota, the giant massive terminal moraine that's a continuation of Long Island all the way straight up through to Cape Cod, the glacial soils that cover much of the middle part of the country, the existence of the Ogallala Aquifer, all that water that was recharged underground from the runoff of the large ice sheets. Remember the giant lakes that once covered parts of Utah and other states that evaporated away to leave us the giant salt flats like the Great Salt Lake and the Bonneville Salt Flats, the massive carved-out valleys like the Alps and Yosemite, where the motion of the glaciers tore away large portions of the rock. I talked about the rebound of Canada, moving upward as the ice was removed and the asthenosphere began to flow back underneath. I talked about the giant ripple marks that cover parts of the northern plains, where these massive glacial lakes burst dramatically, flooding the surface with enormous amounts of water. I mean, for example, the Great Lakes today represent some of the largest lakes in the world, but they're really just puddles that are left over from the last Ice Age. They were once huge, enormous lakes, much larger, containing a lot more water than they do today.

Now, times of global cooling, like these last Ice Ages, also have very important effects on life and living organisms, on the very evolution of species, both plant and animal. It's tough to live through a time where the temperature drops dramatically and it's really cold. As the global climate would have cooled gradually, regional climate bands would have shifted all around the planet.

In other words, alpine environments would have dropped down to much lower elevations, and zones of temperate forests would have extended to lower latitudes, further from the poles. And this is in areas that weren't entirely covered with ice anyway.

During the recent Ice Ages, cold temperatures also favored the evolution of larger animal sizes, because if you're larger, your heat has to go farther to get to the surface. You just retain your heat much more efficiently.

So, many mammals such as the giant mammoths, and elk, and sloths evolved to be enormous. There were ground sloths that were more than 20 feet high. There were horses whose backs were taller than an adult human. And, interestingly, as the climate warmed, these species no longer survived, because their whole body structure had evolved to deal with very cold temperatures.

The drop in sea levels during the Ice Ages also had very interesting effects on Earth's geology and biology. Suddenly you had enormous expanses of exposed land—again, those layers that weren't covered with ice. And so, you ended up with very increased rates of erosion.

Remember, back in geologic time, I talked about unconformities, places where the geologic record was missing. Well, if you've suddenly dropped the sea level (because you've put all the water up on the land as ice), you erode the surface everywhere. And layers of rock that may have taken tens of millions of years to form now get eroded and are permanently lost.

Interestingly, you've greatly increased the total area of continents. I mentioned earlier that continents actually occupied 39% of the surface of the earth. However, 9% of the surface of the earth, about a quarter of the continents, are currently flooded, are under water because of the rising sea levels. So, 30% of our surface is land, but a large portion exists as continental shelf. Those are the very fertile, shallow marine environments.

Well, during times of glaciers, all those areas are brought above sea level, and so your marine environments are very limited to the steeper continental slopes and tend to be quite narrow zones. These are rougher times for marine organisms, as well as causing a change in global climates, because often, larger continents mean a shift towards monsoon type climates, a large annual alternation between rainy periods and dry periods.

You drop the sea level, you also do something very interesting—you connect continents. The Bering Strait is very shallow; it's about 150 feet deep. Well, the sea level was 400 feet deeper during the time of the last Ice Ages. That means that Eurasia, and North America, and South America were all one continuous landmass.

So, about 20,000 years ago, woolly mammoths that had evolved over in Eurasia crossed over from Siberia into North America, and following soon after were us, humans. Humans came across into North America from Asia about 14,000 years ago. And interestingly, they brought their domesticated dogs with them as well, the first dogs to come into North America.

At the same time in Southeast Asia, the lowering of the sea level there connected Australia with the rest of Asia. The native aboriginals that lived in Australia now came over at that point, were able to walk across, or, at least, move across very shallow regions, in order to get from Southeast Asia into Australia.

Now, all this came to a rapid end 10,000 years ago. Why? Because of these Milankovitch cycles. It was the end of the glacial period; it was the start of the next interglacial. It's just the way the different orbital parameters combined. It's part of that whole intermediate time cycle of global climate change. There was a very rapid rise in temperature and the start of a warm stable period. Interestingly, it also coincided, and not coincidentally, with the start of human civilization. I'm going to talk more about this in Lecture Forty-One.

But first, I need to explain the reasons that we have very long time scale changes. I mean even longer than the Milankovitch cycles. I'm going to do that in the next lecture, and I also need to explain some of the reasons that we have very short time scale climate changes, and I'm going to do that in Lecture Forty.

Lecture Thirty-Nine
Long-Term Climate Change

Scope:

Throughout the geologic record, there is a strong correlation between global temperatures and the levels of carbon dioxide in the atmosphere. This is because carbon dioxide (and water vapor, ozone, methane, and others) is a greenhouse gas. The greenhouse effect, whereby radiation from Earth's surface is absorbed and held in the atmosphere by greenhouse gases, is essential for the existence of life on Earth. However, if carbon levels in the atmosphere get too high, Earth's surface can overheat. Carbon dioxide levels are kept low through the sequestration of carbon by the oceans and by photosynthetic biomass. Earth has gone through many periods of extended freezing, some of the most notable being the "Snowball Earth" conditions that occasionally existed between 800 and 600 million years ago when carbon dioxide was significantly sequestered into marine sediments. The long time-scale variations in climate are largely controlled by plate tectonics. When large plate collisions occur, erosion rates increase along with the growth of mountains. The increased erosion pulls more carbon out of the atmosphere in the form of carbonic acid, which reacts with mountain rocks to make by-products like carbonate molecules. These molecules get carried by streams to the oceans, where they become stored on the ocean seafloor as rocks like limestone. The global cooling that has occurred over the past 50 million years is largely due to the formation of the Himalayan Mountains.

Outline

I. The discussion of long-term climate change.

    A. The insulating effect of the greenhouse gases—water vapor, carbon dioxide, and methane—makes our planet livable; however a runaway greenhouse effect is not good.

    B. Global climates over Earth's history track closely with changes in the greenhouse gases.

    C. Milankovitch cycles changed climate patterns on Earth over time scales of tens to hundreds of thousands of years.

II. Plate tectonics drive climate change over tens to hundreds of millions of years. Mantle convection and plate tectonics, through interactions with the process of weathering and with the carbon cycle (and therefore with the greenhouse effect) have shaped the evolution of life on our planet.

    A. One hundred million years ago (during the time of the dinosaurs) there were very high global temperatures.

        1. The continents were flooded and there was no ice at the ice caps.

        2. It was very hot and the amount of carbon dioxide in our atmosphere was more than double the levels that it is today.

    B. If you look over all the different time scales from shortest to longest, you see a very strong correlation between Earth's global temperature and the amount of atmospheric carbon dioxide.

    C. In order to understand this, we need to examine how the greenhouse effect works, and then how the whole carbon cycle works.

    D. Water vapor, carbon dioxide, ozone, and methane in our troposphere form the basis of our greenhouse effect.

        1. This greenhouse effect traps heat near Earth's surface.

        2. It keeps the land warm enough to support liquid water.

        3. It is absolutely requisite for the existence of life as we know it.

III. How does the greenhouse effect work? What happens?

    A. Sun's radiation, as it reaches Earth's surface, is peaked in the visible range and bleeds over into the ultraviolet range.

        1. This is why your eyes have evolved to see radiation at frequencies corresponding to what we call "visible light."

        2. The only difference between visible light and other forms of electromagnetic radiation is simply the wavelength of that radiation.

    B. As that radiation from the sun hits the surface of Earth, it goes through a complicated series of reflections, absorptions, and re-emissions.

        1. Thirty percent gets directly reflected back out into space.

2. Twenty percent of that is absorbed by the atmosphere and is then re-radiated right back out into space again.

C. When that sunlight and ultraviolet radiation hits the atoms of Earth's surface, the atoms absorb the energy. The electrons jump to a higher level of activity, called a "quantum leap."

D. The radiated energy gets momentarily stored within the atom as a higher level of kinetic energy, but this state is not stable. The electron drops back down to its ground level, re-emitting that energy in the form of new electromagnetic radiation.

E. The radiation that gets emitted by Earth's surface is not in the same form as the radiation that it received. The amount of energy is the same, but it takes a different form because the energy that gets re-radiated is determined by the surface temperature.
   1. Why? The Earth's surface is colder than the sun's: about 14°C, compared to 6000°C.
   2. The Earth takes the sunlight's energy and re-emits it as infrared, which is at a longer wavelength than visible and ultraviolet radiation.

F. Greenhouse gases in the atmosphere preferentially absorb infrared radiation, not visible and ultraviolet.
   1. Sunlight coming directly from the sun passes right through these greenhouse gases and hits the surface of the earth.
   2. When the earth re-emits it as infrared, the atmosphere absorbs this energy because it is at a frequency that corresponds to the atomic structures of these molecules.
   3. Once the atmosphere absorbs that radiation, it re-emits it right away. Some of it heads off into space and is lost, but some of it goes downward and is re-absorbed by Earth's surface.
   4. That infrared energy begins a series of bounces back and forth between the surface and the troposphere, warming the atmosphere and the surface of the planet and making it livable.

G. The mean temperature of Earth's surface is currently 14°C, but this changes significantly over geologic time so that the energy balance is zero: The amount of sun's energy received at Earth's surface is balanced by the amount leaving it.

IV. The greenhouse process occurs primarily because of the existence of water vapor, carbon dioxide, and other greenhouse gases.

    A. Water vapor is much less plentiful than carbon dioxide but more important as a greenhouse gas because of how efficiently it absorbs infrared radiation.

    B. Ozone operates differently because it is primarily located much higher, in the stratosphere, not in the lower troposphere.

    C. Methane is 21 times more efficient at absorbing infrared radiation than carbon dioxide, but it is not as significant because there is much less of it in the atmosphere.

V. One potentially dangerous aspect of climate is that several factors can lead to a runaway greenhouse effect.

    A. Increasing temperature actually increases the greenhouse effect through a set of positive feedbacks, which then increases the temperature even more.

    B. When climates warm, the amount of surface ice is reduced, ice caps shrink, and the sea ice at the North Pole is reduced.

    C. This decreases the planet's reflectance, a quantity we call the "albedo."

    D. More sunlight is then absorbed by the surface and re-emitted in the infrared range to be absorbed by the atmosphere, which makes the atmosphere warmer and melts more ice.

    E. Warmer air holds greater amounts of greenhouse gases. If you make the atmosphere warmer it can hold more water vapor in it, and water vapor is a very efficient greenhouse gas, which then absorbs more infrared radiation, makes the surface warmer, causes more water vapor to form, and so on.

    F. This creates a cycle of positive feedbacks, making global temperatures rise very quickly.

VI. In order to understand why plate tectonics would be responsible for climate change, the carbon cycle needs to be examined.

A. The carbon cycle is a set of interlocking processes that move carbon back and forth between several different reservoirs including the ocean, the biosphere, and the soil.

B. The carbon cycle has two parts: shallow surface and deep Earth.

C. First the shallow part. Earth's atmosphere would be mostly carbon dioxide if it weren't for two factors: oceans and vegetation.

   1. Most surface carbon actually exists as dissolved carbon dioxide in the oceans: 40,000 billion tons of carbon. Carbon also exists in marine life and in organic sediments on the ocean floor.

   2. The pathways between these different carbon reservoirs turn out to be very complex. Every year about 120 billion tons of carbon are moved back and forth between the atmosphere and the entire vegetation/soil system.

   3. Humans are adding about 7 billion tons of carbon to the atmosphere every year. That is enough to upset the equilibrium.

   4. Some of the added carbon is being absorbed by vegetation. Forests are getting stronger and healthier.

   5. Most of the rest is going into the atmosphere and raising global temperatures through an increase in greenhouse gases.

D. Over geologic time the deep Earth interacts with the carbon cycle both by adding and subtracting carbon from it.

   1. Carbon from deep Earth gets added to the atmosphere primarily in the form of carbon dioxide emitted from volcanoes.

   2. Carbon gets added into oceans largely through the erosion of limestone and other carbonate-based sedimentary rocks.

   3. Carbon gets removed from the ocean through sedimentation, through the production of calcium carbonate rocks, through compaction of material at accretionary wedges during subduction, or through subduction of the sediments themselves when the seafloor subducts into a trench.

VII. Carbon also gets removed from the atmosphere through weathering during dissolution.

    A. When carbon dioxide combines with water, it forms carbonic acid. This dissolves rock, and the carbon goes into bicarbonate molecules that are washed down streams and into the ocean.

    B. In the oceans the bicarbonate molecules combine with materials like calcium to make reservoirs of limestone.

    C. This is why we don't have 95% carbon dioxide in our atmosphere. Our carbon has largely been locked away in limestone.

    D. Now we can begin to discuss how plate tectonics affects global climates.

        1. About 150–180 million years ago the climate was steadily warming. That was a period of increased hot spot volcanism.

        2. This period of intense volcanism may have put enough carbon dioxide into the atmosphere to start one of the positive-feedback loops of global warming.

        3. Several times in the past half-billion years were periods where global climates cooled considerably, and each of these corresponds to a period of continental collisions. As mountains rise, weathering increases. Carbon goes into bicarbonate ions which go into rivers, then oceans, then into the shells of marine animals and then onto the bottom of the seafloor where they become carbonate rocks like limestone.

        4. Increased growth in mountains has meant an increased removal of carbon dioxide from the atmosphere and an increased cooling of the entire planet.

        5. If it weren't for the collision of India into Asia and the formation of the Himalayas, our global climate would probably never have cooled to the point of its having the large Ice Ages that it now has on a regular cyclical basis, driven by the Milankovitch cycles.

VIII. There is geologic evidence for several periods of extreme glaciation between about 800 and 600 millions years ago.

A. During these Snowball Earth periods, glacial sediments were embedded within icebergs and then deposited onto the ocean floor long distances from where these glaciers first entered the ocean.

B. Several times during this period the climate swung between runaway greenhouses (very hot periods) and periods so cold that entire ocean surfaces were possibly frozen over. It would have been very hard for life forms and it is probably not a coincidence that there is no fossil evidence for multicellular life until after this time.

C. These large changes in climate may have been possible because of the lack of multicellular life; particularly because of the lack of worms and other creatures that live in ocean sediments.

    1. Worms churn up marine sediments, preventing methane and carbon dioxide from getting locked away within them.

    2. Once worms evolved, greenhouse gases may have been better kept in circulation.

    3. Although it seems as though humans are working hard to become the planet's most significant agent of geologic change, we have a long way to go before we attain the same level of impact as the worm.

D. Plate tectonics drives long-time-scale climate change primarily by changing weathering rates.

E. Milankovitch cycles drive intermediate-time-scale climate change by slightly changing the amount and distribution of sunlight.

Recommended Reading:

Alley, *The Two-Mile Time Machine: Ice Cores, Abrupt Climate Change, and Our Future.*

Hoffman and Schrag, *"Snowball Earth,"* *Scientific American,* January 2000.

Questions to Consider:

1. The global temperature changes during the past 500 thousand years correlate very well with carbon dioxide changes in the atmosphere, yet the amount and distribution of sunlight Earth

receives does not directly affect CO_2. Then why are they correlated?

2. Will global forests become more or less healthy during periods of extreme continental collisions?

Lecture Thirty-Nine—Transcript
Long-Term Climate Change

Welcome. In this lecture, I want to talk about long-term climate change. Now, I've sometimes wondered what it would be like to ride in a spaceship, hurtling through the frozen vacuum of outer space in a tin can, separated from instant death by a thin, insulating wall. And then I remember—oh, yeah, I am! Spaceship Earth has just such a protective thin insulating wall—the atmosphere.

Okay, it's not as thin as that of a spaceship, but it's equally as insulating, and that insulation comes from the greenhouse effect. Remove the atmosphere (like on a smaller planetary body like our own moon) and the temperature would be that of the emptiness of space, $-270°C$ ($-450°F$). It's the greenhouse gases in our atmosphere—which are water vapor, carbon dioxide, and methane—that make our planet livable.

Now, people sometimes talk about the greenhouse effect in a disparaging way. This is very mistaken. It's one very important reason that we're alive. Of course, you can have too much of a good thing, and a runaway greenhouse effect would not be very good. As a result, global climates over Earth's history track very closely with changes in the greenhouse gases. And it's the changes in climate and the greenhouse effect that occur over very long time periods that I want to talk about in this lecture.

Now, in the last lecture I talked about the Milankovitch cycles, the cycles of changing Earth's orbital parameters, and how these changed climate patterns on Earth over time scales of tens to hundreds of thousands of years. But what drives climate change over tens to hundreds of millions of years? Well, the answer, interestingly, is plate tectonics.

In a very direct way, mantle convection and plate tectonics (the processes from the first half of this course, which, maybe, you thought we were done with), through interactions with the process of weathering and with the carbon cycle—and, therefore, with the greenhouse effect—end up shaping the evolution of life on our planet.

For instance, 100 million years ago, the time of the dinosaurs, there were very high global temperatures. The continents were flooded;

there was no ice at the ice caps. It was a very hot climate, and accordingly, the amount of carbon dioxide in our atmosphere was more than double the levels that it is today.

If you go back farther in time, you see even greater climate changes. Global climates have changed enormous amounts over tens to hundreds of millions of years, and the carbon dioxide has varied correspondingly. We even see times when the entire ocean surfaces were frozen over, and there was much less carbon dioxide at that time. If you look over all the different time scales, from short time scales to the very longest time scales, you see a very strong correlation between Earth's global temperature and the amount of atmospheric carbon dioxide.

For instance, let's look at changes in temperature and carbon dioxide over the last 450,000 years. These are the oscillations I talked about in the last lecture, the patterns of Ice Ages and interglacials. There are two curves here. One is temperature; the other is carbon dioxide. And you can see that the pattern of interglacials and long Ice Ages shows up very clearly in both of these curves. Extend this to the smaller cycles or the larger cycles that this period exists as only a small blip in, and again, the pattern, the connection between temperature and carbon dioxide, is very strong.

Well, to figure this out, we need to examine first how the greenhouse effect works. And then we have to look, in order to figure that out, at how the whole carbon cycle works. We need to see why the levels of carbon dioxide in our atmosphere change. Now, the presence of water vapor, carbon dioxide, ozone, and methane in our troposphere form our greenhouse effect. They trap heat near Earth's surface. They keep the land warm enough to be able to support liquid water and, therefore, warm enough for us. So, therefore, they are absolutely requisite for the existence of life as we know it on the earth.

Well, what happens? The sun's radiation, as it reaches Earth's surface, is peaked in the visible range and actually bleeds over strongly into the ultraviolet range. By the way, this is why your eyes have evolved to see radiation at frequencies corresponding to what we call "visible light." The only difference between visible light and other forms of electromagnetic radiation (like radio waves, x-rays, microwaves, etc.) is simply the wavelength of that radiation. We've got a whole lot of visible light coming from the sun, it's what our eyes have evolved to be able to see with.

But here's the interesting point: As that radiation from the sun hits the surface of the earth, it starts going though a very complicated series of reflections, absorptions, and re-emissions. Well, first of all, as the sunlight reaches the earth, some part of it never gets to the surface. Actually, about 30% gets directly reflected back out into space. About 20% of that is absorbed by the atmosphere and is then re-radiated right back out into space again.

Well, when that sunlight and ultraviolet radiation hits the atoms of Earth's surface, a very interesting process occurs: The atoms absorb the radiation energy. I talked about this previously. What happens is that the electrons actually jump to a higher level of activity. This is called a "quantum leap."

That energy gets momentarily stored within the atom as a higher level of energy, but it's not stable, and it doesn't stay there. That electron will drop back down to its ground level, re-emitting that energy in the form of new electromagnetic radiation.

But here's the trick: The radiation that gets emitted by Earth's surface is not the radiation that it received. The energy is the same, but it takes a different form. And the reason is that the energy that gets reradiated is determined by its temperature.

So, the sunlight comes in, it hits Earth's surface largely as visible and ultraviolet radiation, fairly high frequency energy, gets absorbed by the earth's surface and is re-emitted at infrared wavelengths.

Why is this? It's quite simple: Earth's surface is a lot colder than the sun's. The sun's surface is about 6,000°C. The earth's surface is on average about 14°C. I mean, if you were heated up to 6,000 degrees, you would be emitting visible and ultraviolet radiation as well. But the earth takes that energy and re-emits it at infrared, which is a longer wavelength and is a lower form of energy.

But here is where our greenhouse effect works. The greenhouse gases, mostly water vapor, and carbon dioxide, and methane, preferentially absorb infrared and not visible and ultraviolet. So, when the sunlight came directly from the sun, it just passed right through these greenhouse gases and hit the surface of the earth. When the earth re-emitted it as infrared, the atmosphere now absorbs that energy because it's at a frequency that corresponds to the atomic structures of these molecules.

Now, once the atmosphere absorbs that radiation, some of it heads off into space and is lost, but some of it then gets re-emitted and goes back towards the surface—re-absorbed by the earth's surface, re-emitted, re-absorbed by the atmosphere—and that infrared energy begins a series of bounces back and forth between the surface and the troposphere, warming the atmosphere and the surface of the planet and making it livable.

Here's an important aspect, though. For each of these systems, the space, the atmosphere, and Earth's surface, the energy balances to zero. There's an equilibrium. There's no net increase in energy. I mean, a vast amount of energy from the sun is reaching our planet's surface, but it doesn't dramatically increase in temperature. Why? Because the same amount of energy that it receives leaves again. And the same holds true for the atmosphere. However, this equilibrium is not a coincidence. How does this equilibrium happen? It's very simple. The temperature of Earth's surface changes to make this happen.

Remember, I just said that hotter objects radiate more energy. Well, the mean temperature of Earth's surface is 14°C currently, but that number changes significantly over geologic time. If there's more sunlight, or more greenhouse gases, the temperature of Earth's surface will begin to increase. And that means it's going to re-radiate more energy back out, because it's hotter. It will then maintain that equilibrium again. In the process, the temperature in the atmosphere and at the surface may change, but they will always change accordingly so that the balance comes back into zero. The same amount that goes in is the amount that goes out.

Now, this process of having a greenhouse occurs primarily, as I said, because of the existence of water vapor and carbon dioxide, and there are other materials that are important as well. Water vapor is actually much less plentiful than carbon dioxide. Even though its composition occupies a smaller volume of Earth's atmosphere, it's still twice as important as carbon dioxide as a greenhouse gas. It's just how the molecule works, how efficiently it absorbs infrared radiation.

Ozone also turns out to be quite important, but the process operates differently, because ozone is primarily located much higher up, 25 kilometers above the surface in the stratosphere, not in the lower troposphere.

Methane turns out to be an enormously efficient greenhouse gas, 21 times more efficient at absorbing infrared radiation than carbon dioxide is, but it's usually not significant, because there's usually just very little of it in the atmosphere. However, because of our appetite for hamburgers and an increase in ranching and cattle, there is a significant increase in methane, and this is having an important influence on the whole greenhouse effect.

Now, we have to be careful with climate, because several factors can lead to what we call a "runaway greenhouse scenario." In other words, you can have certain factors that will cause the whole process to occur more rapidly. For instance, increasing temperature actually increases the greenhouse effect, which increases the temperature through a set of positive feedbacks.

For instance, let's say climates warm. Well, the amount of surface ice is reduced—in other words, the ice caps shrink. The ice sea at the North Pole is reduced. This decreases the planet's reflectance, a quantity we call the "albedo." More sunlight is now absorbed by the surface and then re-emitted in the infrared range to be absorbed by the atmosphere again, which makes the atmosphere warmer, which melts more ice, which causes more energy to go into infrared. And you can see this positive feedback cycle begin.

Here's another very important factor: If you have warmer air, it actually holds greater amounts of greenhouse gases. Remember that back in my discussion of weather, I said that when air was warmer, it held more water vapor in it, and then when you cooled the air, you cooled the atmosphere, it caused that water to condense and precipitate back out.

Well, if you make the atmosphere warmer, it can hold more water vapor in it. And water vapor is a very efficient greenhouse gas, which then absorbs more infrared radiation, which makes the surface warmer, which causes more water vapor to form and be absorbed in the atmosphere.

And again, you can run into this cycle of positive feedbacks, making a climate go very warm, very quickly. And incidentally, the times in Earth's past where we've seen the temperature increase very quickly over very short time periods, such as the start of these interglacial periods, is probably some form (at a smaller scale) of a runaway greenhouse effect.

Now, I should say, the term "greenhouse effect" is a little bit of a misnomer. It turns out that greenhouses don't actually stay warm because of the greenhouse effect. What happens, more efficiently, is that simply having an enclosed space with glass windows keeps the air that gets warm from the ground and the sun from escaping back out. Nonetheless, we're stuck with that term, "greenhouse effect," and it is very important in our atmosphere.

Well, so far I still haven't explained why plate tectonics would be responsible for climate change over very long time scales. Now, in order to do this I have to explain one more thing. I have to talk about the carbon cycle.

The carbon cycle is a set of interlocking processes that move carbon in between several different reservoirs, which are the atmosphere, the ocean, the biosphere, and soil. Now, as with the water cycle, which we talked about earlier, the carbon cycle has two parts. It has a shallow surface part and a deep Earth part.

Let me talk about the shallow part first. Earth's atmosphere would likely be mostly carbon dioxide if it weren't for a couple factors. Now, if you look at Mars and Venus, our neighbor planets, they have atmospheres that are 95% carbon dioxide, and Earth's atmosphere was probably 95% carbon dioxide as well, early on in its history. But we have oceans, and we have vegetation, and these processes and the existence of these reservoirs pulls the carbon dioxide out of the atmosphere, and that has had a dramatic change on the whole history of our planet.

Most surface carbon actually exists as dissolved carbon dioxide in the oceans, about 40,000 gigatons of carbon. That's 40,000 billion tons of carbon. Compare that to the amount of carbon that is in soils, which is 1,600 gigatons of carbon, the amount in the atmosphere, which is 750 gigatons of carbon, or the amount that's in all the vegetation, which is only 600 gigatons of carbon. The oceans, at 40,000 gigatons, dwarf all of these.

And within the oceans, the carbon isn't just in the form of dissolved carbon dioxide in the water. It also exists in marine life. It also exists in organic sediments on the ocean seafloor. And the pathways between these different carbon reservoirs turn out to be very complex.

But again, they are essentially in an equilibrium. Every year, about 120 gigatons (120 billion tons) of carbon is moved back and forth between the atmosphere and the whole vegetation soil system. About 90 gigatons of carbon cycles between the atmosphere and the ocean. They're constantly moving back and forth, largely in the form of carbon dioxide.

And here's the interesting part. How much are we adding? Well, through burning of fossil fuels and also, interestingly, through the production of cement, which is important, we add about 7 gigatons of carbon to the atmosphere every year. Now, this is small in comparison to the other numbers. However, because we're starting with a system that's in equilibrium, that's enough to upset that equilibrium. Now, some of the carbon that we've been adding is being absorbed by vegetation. The forests that exist around the world are getting much stronger and healthier, because they're breathing more easily. There's more carbon dioxide.

The rest, however, is just going into the atmosphere, raising global temperatures through an increase in greenhouse gases. Interestingly, with the vegetation, it's not clear how long the forests can actually continue to absorb that part of the carbon we're releasing. At some point, that buffer may fill up, and temperatures may rise more rapidly, given the same amount of carbon that's being added each year.

However, over geologic time, we have the deep Earth interacting with that carbon cycle as well, and it does this by both adding and subtracting carbon from it. Now, carbon from the deep Earth gets added to the atmosphere primarily as carbon dioxide emitted from volcanoes. If you have a period of extended volcanism, you get a global increase in temperature. You add a lot of carbon dioxide up into the atmosphere.

Carbon gets added into the oceans largely through the erosion of limestone and other carbonate based sedimentary rocks. However, that carbon usually ends up leading to increased rates of the solution of carbonates out of the ocean and then onto the seafloor, and so it's not going into an increased level of carbon dioxide in the atmosphere.

Carbon gets removed from the oceans in several ways: through sedimentation, through the production of these calcium carbonate rocks like limestones, through the compaction of this material at accretionary wedges during subduction, or through the actual subduction of the sediments themselves down into the mantle, when the ocean seafloor subducts beneath a continent. Carbon also gets removed from the atmosphere through weathering during dissolution, and this is the direction that we need to go in at this point.

When carbon dioxide combines with water, it naturally forms carbonic acid. I talked about this already when I talked about weathering. This is the foundation of the whole process of chemical weathering, of dissolving rocks. The carbonic acid combines with rocks, it dissolves out those rocks, and the carbon goes into bicarbonate molecules that are washed through streams and down into the ocean.

These oceans can take those bicarbonate molecules, they combine with materials like calcium to make reservoirs of limestone, and as a result, the layers of limestone that exist not only underground but across the surface of continents, as platform sediments, represent an incredibly vast reservoir of carbon. In fact, there may be more than 1,000 times as much carbon locked away in rocks like limestone than in the whole rest of the surface carbon cycle combined.

This material, this carbon, was once dissolved in the ocean, and it's now sequestered away, out of the carbon cycle. That's why we don't have 95% carbon dioxide in our atmosphere. Our carbon dioxide, our carbon, has all been locked away in limestone.

And now you can begin to understand how plate tectonics will affect global climates. What I'm going to do is give several examples, using a figure of global temperature change over the past 540 million years, since the start of the Cambrian. For temperature, as a proxy, this figure uses O_{18} isotopes. Remember how this worked? High amounts of O_{18}, the isotope O_{18}, corresponds to cold times, because ice is mostly O_{16}, and you remove the water with O_{16} out of the oceans, and within the oceans, the relative amounts of O_{18}, what we call ΔO_{18}, the change in O_{18}, increases. That's what gets locked away in the marine shells.

Now, a question: Why doesn't this curve go back farther than the Cambrian? Well, it turns out that O_{18}, as I mentioned previously, is taken from ocean shells. And before the Cambrian, there were no shells; there were no hard-bodied fossils of any sort. And so, we can't use the O_{18} record to go back beyond 540 million years.

Well, if we look back, what do we see? During about 150 to 80 million years ago was a period when the climate was steadily warming. Well, what was going on in terms of tectonics? Well, that was a period of increased hot spot volcanism.

Remember that I talked about this back in Lecture Sixteen: you had hot spot activity which was triggered by the increased amount of subduction that followed the breakup of Pangaea. All these slabs sank down into the earth, pushed hot rock aside at the base of the mantle, came up as hot spots, and erupted at the surface as several large outpourings of lava. Remember, that was also connected to the lack in magnetic field reversals that we saw in the ocean seafloor. This period of intense volcanism could have put enough carbon dioxide into the atmosphere to essentially start one of these positive feedback loops of global warming.

You go back beyond that, and several times during the past 50 million years we've had periods where global climates have cooled considerably—enormous amounts, very large changes in temperature. And interestingly, each of these corresponds to a period of continental collisions and mountain building. Well, what happens? As the mountains rise up, as we previously talked about, weathering increases. Weathering hates mountains. Mechanically, chemically, the mountains get torn away quickly. So, carbonic acid dissolves away the mountain rocks—more mountains, more weathering. The process sucks the carbon dioxide right out of the atmosphere.

The carbon goes into bicarbonate ions, they go into rivers, they go into the ocean, they get consumed by marine animals, they go into the shells of the animals or are chemically deposited on the bottom of the seafloor, and they become carbonate rocks like limestone. In other words, making mountains removes carbon from the atmosphere and puts it onto the seafloor.

So, if you look back 480 million years ago, there's a period of cooling that we see in this figure. Well, what happened then? Remember my lecture on the formation of North America? That was the time of the enormous Taconic orogeny, when the whole Taconic mountain system along eastern North America was forming. Then 350 million years ago. This corresponds to an important episode of the formation of Pangaea, when part of that was the formation of the Allegheny and the Ouachita mountains—again, more mountain building along the east coast of North America.

And 280 million years ago was another period of global cooling. That's when the enormous Appalachian Mountains were formed, the Himalayas of their day, a massive, giant plateau of rocks. Enormous weathering would have pulled the carbon dioxide out of the atmosphere and washed it into the ocean.

Most recently, if you look at this figure, you see the cooling has been quite dramatic over the past 80 million years. Now, remember, I said 100 million years ago, when the dinosaurs were ruling the earth, we had really high temperatures, a lot of carbon dioxide, warm weather.

Well, what happened 80 to 60 million years ago? That was the beginning of the Himalayas. India and the Indian Plate began to crash into Asia, sweeping up various island arcs and micro-continent fragments, smashing them all into Asia, and forming the enormous plateau that we see there today. The increased growth in mountains, which has continued today, has meant an increased removal of carbon dioxide from the atmosphere and an increased cooling of our whole planet.

So, not only is India responsible for all the big earthquakes that keep hitting China, and the destruction from large earthquakes and tsunamis like what happened in Sumatra in 2004, India is now also to blame for all the Ice Ages we've been having for the past millions of years. If it weren't for the collision of India into Asia, and the formation of the Himalayas, our global climate would probably never have cooled to the point of its having the large Ice Ages that it now has on a regular cyclical basis, driven by the Milankovitch cycles.

Now, we don't have a good record of the change in temperature before about 540 million years. As I said, we just don't have any fossils from back then. But there is geologic evidence for several periods of extreme glaciation between about 800 and 600 million years ago. I talked about this previously.

This was a phenomenon called "Snowball Earth," where we see the glacial sediments that were embedded within icebergs covering the oceans, covering the ocean floor, long distances away from the locations where these glaciers were entering the ocean. It seems that several times during this period of 800 to 600 million years ago, the climate swung severely between very, very hot periods—runaway greenhouses—and periods so cold that possibly the entire ocean surfaces were frozen over.

Now, this obviously would have been very hard on life, and it's probably not a coincidence that multicellular life didn't get started until after this time. But interestingly, it may be actually because of the lack of multicellular life—because of the lack of worms and other creatures that live in ocean sediments—that these large changes in climate even occurred.

Why is this? Well, worms in ocean sediments keep the gases and other sediments all churned up. They help prevent methane and carbon dioxide from getting locked away. Perhaps, without worms, the level of the atmospheric greenhouse gases could run down and could drop precipitously; it could drop so much that the oceans actually froze.

Perhaps this occurred until some sort of catastrophic event, maybe a volcanic event. It might have triggered the release of enough methane from ocean sediments to kick the climate back into a runaway greenhouse, causing a global warming, melting more methane, causing more release of greenhouse gases, and sending the climate careening all the way to the other spectrum of being enormously hot with all the ice melting.

Now, once worms evolved, this no longer happened; the greenhouse gases were kept in circulation. Ironically, though it seems like humans are working very hard to become the planet's most significant agent of geologic change, which I'll talk about more later on in the course, we actually have a long ways to go before we attain the same honor in the climate change hall of fame as the worm.

Well, plate tectonics drives long scale climate change largely by changing weathering rates. Milankovitch cycles drive intermediate scale climate change by slightly changing the amount and distribution of sunlight. And in the next lecture, I will start to look at short term climate change, all those interesting short spikes in the temperature charts we saw—very rapid changes in temperature. And when we do, things will really start to get interesting.

Lecture Forty
Short-Term Climate Change

Scope:

Climates change on shorter time scales than the tens of millions of years of plate tectonics and the tens of thousands of years of the Milankovitch cycles. Circulation of water in the ocean is responsible for carrying heat around the globe, and when these patterns change, regional climates change. For instance, global warming can flood the North Atlantic with freshwater, shutting off the Gulf Stream, and thus chilling eastern Canada, Greenland, and Western Europe. A reversal of equatorial currents in the Pacific (called El Niño) can bring warm waters to western South America, shutting off coastal circulation and killing plankton and fish communities. Changes in radiation output from the sun also change Earth's climate: There is a correlation between sunspot activity, indicative of solar convection patterns, and Earth's global temperature. In addition, sudden changes in climate can occur from volcanic eruptions. Dust and aerosols ejected from volcanoes can block incoming solar radiation, lowering global temperatures. However, volcanoes also have a long-term warming effect through the emission of carbon dioxide into the atmosphere.

Outline

I. Climate can change rapidly over short time scales and has always done this long before people came into the picture.

   A. Several factors affect climate at shorter time scales and cause rapid, instantaneous changes that have a tremendous effect on humans. These include variations in sunlight, ocean current fluctuations, and the eruptions of volcanoes.

   B. There are many different interactions between the geosphere, biosphere, hydrosphere, cryosphere, atmosphere, and solar input. All of these work together to affect change.

   C. Identifying causes for climate events is still a very speculative field of research.

      1. There is not necessarily a direct causative effect between correlated events.

2. There may be two or more causes for a certain event.

3. There may be events occurring due to causes and mechanisms not yet identified.

II. Our whole surface geology system is driven by the sun, so changes in sunlight ("insolation") have tremendous effect on Earth's climate.

A. There is one mechanism for going back in the past and figuring out the sun's historic output: sunspots.

B. The sun is an average sized star with fusion occurring at the core, where energy gets released. Three hundred forty trillion trillion trillion protons fuse to form helium nuclei every second. The process of nuclear fusion destroys a small amount of mass, which gets converted into electromagnetic energy, and some of that energy heats the surface of Earth.

C. Inside the sun the temperature is so great that heat gets transmitted out of its core by radiation, which drives a large convection zone on top of it. In the convection zone, solar plasma convects in tall, narrow columns with a bumpy, knobby appearance at the top.

D. Although the sun doesn't really have a surface, we call the photosphere the sun's surface because it is the boundary for visible light.

E. The sun's photosphere has a temperature of 6000 degrees Kelvin (K) that determines its radiation. It is primarily peaked in the visible spectrum of electromagnetic radiation but also contains large amounts of ultraviolet and infrared radiation.

F. The bumpy surface of the sun has some unusual characteristics. It appears to have spots on it.

1. The number of sunspots at any given time can range from none to a couple hundred.

2. Sunspots are regions of very strong magnetic fields that partially inhibit the sun's convection. They are slightly cooler and therefore appear darker to us.

3. Sunspots occur at times when the sun's photosphere is slightly hotter elsewhere and emitting slightly more energy.

G. Over the last few decades we have seen a strong correlation between the number of sunspots and the amount of radiation the sun emits.

H. The sunlight we receive varies only between 1365.5–1366.5 watts per square-meter. However that is enough to affect Earth's climate.

I. If you look back over the last few hundred years at the number of sunspots on the sun's surface, you see long-term variations.
   1. In the 1600s and 1700s there were hardly any sunspots. We call this time the "Maunder minimum"—a minimum in solar output received at the earth.
   2. The number of sunspots peaked from 1900 to 1950. The increase in global temperatures in the first part of the 20th century is partly due to this increase in solar output.
   3. Since 1950, the sun's surface has cooled off a little bit, and the number of sunspots has decreased.

J. The record of solar insolation only goes back to the 1600s because before then there were no telescopes, so we don't know how the sun's output fluctuates over longer timescales.

K. The Little Ice Age between 1550 and 1850 corresponds to a decrease in solar output (the Maunder minimum and the Dalton minimum).
   1. Sea ice around Iceland that had been nonexistent during the warmer Middle Ages clogged the seas for weeks starting in about 1600, and peaked by 1800.
   2. In 1690 it was so cold that Eskimos landed in Scotland and many Scots moved southward, immigrating to Northern Ireland and setting the stage for future political unrest there.

L. Some historians attribute the explosion of culture that happened during the Baroque period to the fact that it was cold and people spent more time indoors.

M. We see climate change at the end of the Little Ice Age (approximately 1840), when the increased warming led to a period of high humidity in Europe. This led to the potato blight in Ireland that killed millions and caused the migration of millions of others to America.

III. Another geologic cause of sudden climate change is ocean circulation.

A. We have a similar problem for ocean circulation that we have for sunspots: no physical record. We only have human historical accounts.

B. Satellite monitoring provides us with a direct record of surface temperatures and patterns of motion, but this is very recent.

C. Oceans are huge reservoirs of heat. Ocean currents, therefore, carry heat all about the surface to various parts of the world. If there is a change in the patterns of ocean currents, there will be significant changes in regional climates.

D. One of the most important patterns of ocean circulation changes is the El Niño effect. El Niño is often followed by La Niña conditions. The whole cycle is called the El Niño Southern Oscillation (ENSO).

    1. With El Niño, warm Pacific equatorial currents that usually go from east to west change their direction to flow west to east.

    2. Warm water arrived on the coasts of Central and South American with deleterious effects on fishing yields there.

    3. El Niño is also a coupled ocean atmosphere system, with large changes in atmospheric circulation patterns.

    4. When El Niño affects ocean currents, it produces extensive droughts on the eastern side of the Pacific. On the western side there are large typhoons and thunderstorms.

    5. El Niño is followed by a La Niña condition that involves warm waters moving farther west. The system acts like a big spring, with waters flowing one way with El Niño, then reversing and flowing back again with La Niña.

    6. This large-scale oscillation of the ocean convection pattern occurs over a short time scale (3–8 years) and is of variable strength.

    7. The ENSO is clearly the strongest of the current ocean oscillation phenomena, but we do not yet know what drives its size, strength or timing. It involves a very complex set of interconnected systems.

E. Another example of changes in ocean oscillation is what occurs when the Gulf Stream is halted. The Gulf Stream does not shut down on a regular basis, but when it does it can have dire consequences for Europe.

    1. The Gulf Stream is an ocean current of warm water from the equatorial Atlantic flowing in through the Gulf of Mexico, up along the east coast of North America, across the top of the Atlantic, toward the North Atlantic, and then sinking down, to return southward at a deeper level.

    2. If there is a sudden flood of fresh cold water from melting glaciers, the whole circulation pattern of thermohaline convection stops and shuts down.

    3. Ironically, you can have a period of global warming shut down the Gulf Stream and end up sending Western Europe into a deep freeze. We think that is what happened during 500–400 B.C.E.

    4. During this time of intense cold, many Europeans moved south to find better places to grow crops. As part of this, the Macedonians moved southward from central Europe and invaded Greece, setting the stage for the Macedonian king Alexander the Great to form the first giant civilization encompassing Europe and parts of the Middle East.

IV. Some of the most dramatic, sudden, and catastrophic changes in climate are due to volcanic eruptions.

    A. Greenhouse gasses such as carbon dioxide and water vapor erupt out of volcanoes and, over long periods of time, cause global warming.

    B. Over short time scales, volcanoes have the opposite effect because fine dust and aerosol particles ejected from the volcano block out sunlight.

    C. This blocking of the sunlight can reduce global temperatures by greatly increasing the reflectance of sunlight by Earth's atmosphere.

D. Aerosols are more important than dust particles because they can last in the atmosphere for months to years.

1. For example, Krakatau erupted in 1883. Sulfate aerosol levels in the atmosphere increased by a factor of five, and remained high for half of a year.

2. Global temperatures were unusually low for three years following the eruption.

3. The atmospheric dust and aerosols cause richly colorful sunsets. Edvard Munch's painting *The Scream* shows a dramatic red, almost violent sunset in the background. He started painting it right after the Krakatau eruption.

E. The French revolution occurred in 1789 because of two volcanic eruptions, the Hekla volcano in Iceland and the Asama volcano in Japan, both of which began in 1783.

1. Tremendous amounts of aerosols that spewed out from these volcanoes decreased temperatures around the globe.

2. The winter before the storming of the Bastille was the coldest of the century. Crops failed and massive starvation occurred.

3. France was only one of about a dozen governments in Europe that collapsed. There was widespread change in the whole socioeconomic situation in Europe, where mass starvation led to significant political unrest.

F. After 1815 there was a large push of Americans heading westward.

1. This was prompted by large crop failures in the eastern United States that were the result of the eruption of Mount Tambora, a volcano in Indonesia.

2. The year after the eruption, 1816, became known as the "year without a summer," and was the coldest year since 1601 (which, interestingly, followed a large volcanic eruption in Peru).

G. The eruption of Mount Tambora in Indonesia was the largest in over a thousand years.

1. It ejected about 100 cubic kilometers of tephra into the atmosphere.

2. Global temperatures dropped rapidly due to all the dust and particulate matter ejected into the atmosphere, and

sent many parts of the world, including Europe, into a deep freeze.

3. Spectacular sunsets were also observed after Tambora's eruption. The paintings of Joseph Mallory Turner feature dramatic sunsets, however his earlier sunsets don't have the same colors in the years before the Tambora eruption.

H. Interestingly, a study in 2007 of more than 500 paintings showed that the sunsets of painters such as Turner, Rembrandt, Reuben, Degas, Copley and Gainsborough, were significantly more colorful in years following a large volcanic eruption.

I. During 1150–1136 B.C.E., Hekla volcano in Iceland had a large eruption. Ashes rained on China for 10 days and some anthropologists have suggested that 90% of the population of Scotland and England died at that time.

J. According to Chinese historian Pan Ku, in 209 B.C.E. great famines killed more than half the population as a result of an eruption in Iceland. Stars weren't seen at night the following year for about 3 months.

K. The Byzantine historian Procopius wrote that in the year 536 C.E. 80% of the population starved to death. Population losses on the order of 80%–90% are hard for us to imagine in today's world, but we can go back over just the last few thousand years and find several times when volcanic eruptions have caused this to happen.

V. However, there was one eruption that may have been more significant than any of the others.

A. Within the climate record there is a period of severe Ice Ages that began about 75,000 years ago and correlates with the start of modern human. If you look at the mitochondrial DNA of *Homo sapiens*, it seems that most living humans evolved from a small set of common ancestors about 75,000 years ago.

C. We think *Homo sapiens* originated in Africa about 200,000 years ago, but we are not descended from all of these people.

D. We may have descended from just a small group of people that lived about 75,000 years ago.

E. Also occurring about 75,000 years ago was the largest eruption in the last 100,000 years: Toba Volcano, in Indonesia. The explosion of this volcano was equivalent to a gigaton of TNT, and ejected 280,000 cubic kilometers of tephra into the air. The caldera from this eruption is 100 kilometers in length; the largest in the world.

F. Some anthropologists propose that the strain on the environment was so severe that most humans died. In a few places like the warm, sheltered rift valleys of Africa a few tribes would have survived, and all modern humans would have descended from those survivors.

G. These sorts of phenomena don't happen often, but they always have the potential of happening again.

Recommended Reading:

Cox, *Climate Crash.*

Linden, *Winds of Change.*

Questions to Consider:

1. How is it that volcanoes can both cause global cooling and global warming?

2. Some short-term climate is global, and some is regional. Explain which of these is more commonly the case for changes in solar insolation, ocean convection changes, and volcanic eruptions?

Lecture Forty—Transcript
Short-Term Climate Change

Welcome. In this lecture, I want to talk about how climate changes over short time scales. Now, there's been a lot of attention paid to the whole topic of human impacts on climate, and I don't want to take anything away from the importance of this question. It's not going to go away, and in fact, it may well be the dominant question that ends up defining the 21st century. But first, we really need to understand what climate does on its own before people come into the picture.

In the last two lectures, we looked at climate variations at timescales from tens of thousands to hundreds of millions of years, but what about change over years, or decades? Well, it turns out that there are several factors that affect climate at these shorter timescales. Variations in sunlight, ocean current fluctuations, eruptions of volcanoes—these different mechanisms cause very rapid, instantaneous changes in climate that have had a tremendous effect on humans.

I've already talked about how complex variations affect the whole climate system. There are many different interactions between the geosphere, the biosphere, the hydrosphere, the cryosphere, the atmosphere, and solar input. All of these work together to affect climate change.

There are certain instances where we can actually look at a particular climate event and identify a possible cause for it, and I'm going to try to do that here. However, I've got to warn you that this is still a very speculative field of research. There is not necessarily a direct causation between events that are correlated. It's possible that there might be two or more causes for a certain climate event. In fact, it's even statistically likely that sometimes there are multiple causes for particular events. Or, even still, there may be many cases where certain climate changes are occurring due to causes and mechanisms that we haven't even identified yet. However, there are some that we've begun to identify, and I'm going to try to outline those here.

First of all, our whole surface geologic system is driven by the sun, so changes in sunlight have a tremendous effect on Earth's climate. We call these changes in "insolation," not "insulation." The insolation is the amount of sunlight we receive on Earth's surface.

These changes turn out to be small, but end up playing very important roles in controlling Earth's climate.

Our understanding of this is relatively new, and there is still a lot of uncertainty. A major part of the problem is that we don't have fossil sunbeams. In other words, there's no record going back in the geologic past of how much sunlight Earth's surface received. But that's not entirely true. We have one mechanism for going back in the past and figuring out sun's temperature, and those are sunspots.

It's a very interesting situation. Our Sun, as I mentioned previously, is an average sized star. The fusion goes on in the core. This is where the energy gets released; you get 340 trillion trillion trillion protons fusing to form helium nuclei every second, and that process of nuclear fusion destroys a small amount of mass that gets converted into energy, and it's that energy that heats the surface of the earth.

Now inside the sun, interestingly, the temperature is so increased that that heat gets transmitted out of the core by radiation, unlike Earth's interior, and that powers (at the top part of the sun) a large convection zone. Here, the solar plasma—essentially, it's just protons stripped of their electrons—convect in tall, narrow columns, and the tops of these convection columns have sort of a bumpy, knobby appearance, and that's what is the sun's photosphere. That's what we see.

The sun doesn't really have a surface; it's just plasma, but we call the photosphere the sun's surface because that's the boundary for visible light. That surface has a temperature of 6,000 degrees Kelvin, and that determines the radiation that leaves the sun. It's primarily peaked in the visible spectrum of electromagnetic radiation but also has large amounts of ultraviolet and infrared.

So, that bumpy surface of the sun has some unusual characteristics. It sometimes appears to have spots on it. That's what we call the sun's "sunspots," and sunspots are tremendously variable. They range from none to a couple hundred at any given time. They're not islands. Remember that the sun doesn't have a surface. They're regions of very strong magnetic field that slightly inhibit the sun's convection, making the surface at that spot a little bit cooler and, therefore, appearing darker to us. But interestingly, the times the sunspots occur are times when the sun's photosphere is slightly hotter elsewhere, emitting slightly more energy.

©2008 The Teaching Company.

If you look at a correlation over the last decades of time, we see a very strong correlation between the number of sunspots that are visible and the amount of radiation the sun is emitting. The amount of sunlight we get varies very slightly. This irradiance varies between 1,365.5 watts per meter squared to 1,366.5 watts per meter squared. It's a very small amount, 1 part in 1,000. However, that's enough to play a significant role in affecting Earth's climate.

So, we have this correlation—more sunspots, more energy—and it varies over short time scales with a very clear period of about 11 years. It may also vary over much longer time scales, and that's a very interesting point. If you look back over the last few hundred years at the number of sunspots that have existed on the surface of the sun at any time, we see long range patterns.

In the 1600s and 1700s, at times there were no sunspots. This is a time we call the "Maunder minimum," a minimum in solar input received at the earth. In the 1800s, there was a period called the "Dalton minimum"—again, a decrease in the number of sunspots and, therefore, the sun's energy.

Interestingly, the number of sunspots sort of peaked going from 1900 to 1950. In fact, the increase in global temperatures on Earth in the first part of the 20th century is partly due to this increase in solar output. Perhaps one-third of the increase in temperature we saw is due to a slight warming on the surface of the sun. However, since 1950, the surface has cooled off a little bit, and the number of sunspots has decreased.

Why does this record only go back to the 1600s? Well, before then, there weren't any telescopes. No one was looking at the sun and counting the sunspots on it, so we have a mechanism that goes back a few hundred years, and that gives us the sense that, indeed, the sun's output does fluctuate—maybe over even long timescales. We just don't know that yet.

We can look back at these times. The Little Ice Age between 1550 and 1850 corresponded to this decrease in solar output, the Maunder minimum and the Dalton minimum, periods of depressed solar activity with fewer sunspots. Sea ice around Iceland, for example, that had been nonexistent during the warmer Middle Ages now clogged the seas for weeks out of the year, starting in about 1600. This actually peaked by 1800, when, for half the year, the sea was clogged and frozen up with ice around Iceland. In 1690, it was so cold that Eskimos actually landed in Scotland, and many Scots moved southward, immigrated to Northern Ireland, setting the stage for political unrest for centuries after. This was a period of cold throughout Europe, and life moved inside at this time. This also saw a great increase in the arts, music, and painting, the whole sort of explosion of culture that happened during the Baroque period. Some historians have attributed it to the fact that it was cold and that people spent more time indoors.

Interestingly, we also see climate change at the end of one of these periods of minimum solar output, at the end of the Little Ice Age in the mid-1800s. In the 1840s, this increased warming led to a period of rain in Europe and this caused the potato blight in Ireland that killed off millions of Irish and caused a huge migration of Irish to America.

Another region where we can point to particular events of sudden change in climate involves ocean circulation, although we have a similar problem for ocean circulation that we have with sunspots and the output of the sun. We don't have a fossil record of ocean circulation changes. We only have historical records and accounts.

Now we have satellite monitoring that provides us with a direct record of surface temperatures and patterns of motion at any given point. But this is very recent, and we're really just beginning to understand the tremendous variations in the transfer of heat between different regions of the ocean.

Remember that I talked about the oceans as being this conveyor belt system that's constantly shifting and changing, moving water throughout the oceans. And because of the very high latent heat of water (remember, the fact that you need to give water a lot of energy to change its temperature), the oceans are huge reservoirs of heat. These currents, therefore, carry heat all about the surface, to various parts of the world, so if you have changes in the patterns of ocean currents, you have significant changes in regional climates.

I already talked about one of the most important patterns of ocean circulation changes, the El Niño effect, which is actually tied into a broader oscillation pattern. El Niño is often followed by La Niña, and it's often called ENSO, for El Niño Southern Oscillation.

©2008 The Teaching Company.

So, what happened with El Nino? We had warm Pacific equatorial currents, that usually go from east to west, changing their direction and flowing west to east. The warm water arrives on the coasts of Central and South America, which had really deleterious effects on fishing yields there. El Niño, however, is also a coupled ocean atmosphere system. At the same time, it involves large changes in atmospheric circulation patterns. So when you have an El Niño affecting ocean currents, you also have extensive droughts on the eastern side of the Pacific. And you get large typhoons and thunderstorms occurring on the western side. El Niño is followed by a La Niña condition, which involves warm waters moving even further west than normal. So, it's almost like a big spring where the waters flow one way for a while and then the water flows back again with La Niña.

This whole process, this oscillation of the large scale ocean convection patterns, occurs over a scale of fairly short times by human standards—about three to eight years—and is of variable strength. We don't quite know what drives that particular period or what causes some ENSO oscillations to be strong and some small. They are clearly the strongest of the current ocean oscillation phenomena, but what exactly drives them we still don't have a direct sense. Perhaps solar output plays a role, but if so, why doesn't ENSO occur at an 11-year cycle?

Part of the problem in trying to figure this out is that what we have is a very complex set of interconnected systems. I like to think of it as a whole group of springs of different shapes and sizes, all connected, like some strange Rube Goldberg contraption, and if you push on it in one place, there's not a direct correlation as to what's going to come out someplace else. We haven't been able to predict that yet. We may at some point, but we're just not there.

I talked about another important example of changes in ocean oscillation, and that's what occurs when the Gulf Stream shuts off. Now, the shut off of the Gulf Stream doesn't occur on a regular basis, but it can have dire consequences for Europe. Let me give you an example. Suppose you have a flood of cold water into the North Atlantic. This cold water you would think of as being heavy, because it's cold, but it's actually buoyant, because it's not salty. It's fresh water, so it doesn't sink down. Remember that the Gulf Stream involved a process where you had warm water from the equatorial

Atlantic flowing in through the Gulf of Mexico, up along the east coast of North America, and across the top of the Atlantic, towards the North Atlantic, and then sinking down again.

Well, if you suddenly flood the top part of the Atlantic with fresh, cold water, that whole pattern of thermohaline circulation ("thermohaline"—temperature and saltiness, those are the two characteristics that affect the density of the water), that circulation of thermohaline convection stops and it shuts down. If the cold water doesn't sink, the Gulf Stream can't move up to replace it.

The fresh water comes from a variety of sources, but usually from melting glaciers or the sudden burst of a glacial dam, and it can serve to shut down the convection in the North Atlantic. Remember that these processes of bursting glacial dams can be quite catastrophic. Remember that I talked about the ripple marks 100 meters apart in North America where water flooded across.

There's another great example in Europe. The English Channel was actually cut by such glacial dams bursting. Actually the English Channel was formed by two such floods, both at the start of one of these interglacial periods, one 425,000 years ago and the other 225,000 years ago.

So, it's ironic. You can have a period of global warming which can shut down the Gulf Stream and end up sending Western Europe into a deep freeze. We actually think that this happened about 500 to 400 years B.C.E., following an extended warming period.

Interestingly, during this time, because Europe was so cold, many Europeans moved southward in order to find greener pastures, better places to grow crops. At that time, 500–400 B.C.E., the Macedonians moved southward from more central Europe, invaded Greece, essentially overran the Greek culture, and set the stage for the Macedonian king Alexander, Alexander the Great, to essentially form the first giant civilization encompassing Europe and parts of the Middle East.

So, I've talked about sunspots. I've talked about ocean circulation. Perhaps some of the most dramatic and sudden changes in climate are due to volcanic eruptions. These can, and have, been the most catastrophic of these climate changes that can significantly altered the course of human history instantaneously.

Now I've already talked a little bit about what comes out of a volcano—the carbon dioxide and the water vapor. Now, these are greenhouse gases. So, over long periods of time, volcanic activity causes, in general, warming. Remember that this is what happened following the whole breakup of Pangaea, where I had slabs sinking into the earth, rock rising up, and lots of hot spot volcanism that caused a lot of the warming that happened during the Cretaceous period.

Well, over short time scales, volcanoes have just the opposite effect, because very fine dust and aerosol particles that are ejected into the atmosphere block out sunlight. It's like putting up an umbrella, a parasol, blocking out sunlight. And this can reduce global temperatures over very rapid timescales by greatly increasing the reflectance of Earth's atmosphere. Sunlight just doesn't get in.

Now, interestingly, the dust is important, but what's actually more important are aerosols, very small tiny droplets of materials (like sulfuric acid droplets), because these can last in the atmosphere for months to years. In fact, their effects can even last longer.

For example, Krakatau erupted in 1883—catastrophic volcano. Following that eruption, sulfate aerosol levels increased by a factor of five. They were five times higher than usual in the atmosphere, and that lasted for about a half a year. However, global temperatures dropped for three years following the volcano. Now, interestingly, there's another effect of having all this little particulate material in the atmosphere: It causes dramatic sunsets. Some of you may know the painting Edvard Munch's *The Scream*, with this dramatic, red, almost violent, sunset in the background. Well, he started painting that right after 1883, right after the Krakatau eruption.

You can trace throughout human history cases where sudden climate events actually have significantly affected human activities. For instance, why did the French Revolution occur in 1789? If you've taken courses in modern European history, you probably have a whole bunch of reasons having to do with a variety of socioeconomic problems developing in Europe. Well, it's much simpler than that: It was volcanoes. In 1783, Hekla volcano in Iceland and Asama volcano in Japan erupted violently. These spewed out tremendous amounts of aerosols that decreased the temperatures all across the globe. In fact, the year before the storming of the Bastille was the coldest winter of the century. Crops failed everywhere. Massive starvation occurred.

It's interesting. There was an American statesman living in France at the time who reported that the sun shone weekly all summer and gave off little heat, and he actually guessed that the cause was due to the Iceland volcano. You can probably guess who that was. It was Ben Franklin, the same guy who had figured out the whole process of mantle convection and had mapped the Gulf Stream.

France fell—the storming of the Bastille happened in 1789—but it was only one of about a dozen governments in Europe that collapsed. It was a widespread change in the whole socioeconomic situation in Europe. Why? You know, the whole thing with the "Let them eat cake," because they didn't have bread. Well, why didn't they have bread? They didn't have bread because the volcano had dropped the temperature and had caused the crops to fail and people were starving. And when people are starving, there is generally political unrest or changes of a variety of types.

Here is another example. After 1815, there's a huge push of Americans heading westward. A huge westward expansion happened at this time. Why? I mean, what would make you pick up your belongings, throw them in the back of some covered wagon and head off into some unknown territory in the western U.S.? Well, you'll do that if you're starving to death, if you don't have any food.

Why didn't people have any food? Because of the eruption of a volcano, Mount Tambora. The year after that eruption, 1816, is known as the "year without a summer." It snowed in New England in the summer. It snowed in June. It was the coldest year since 1601, which, interestingly, followed a volcanic eruption in Peru. The worst famines of the 19th century happened then.

An interesting quirk—the Farmer's Almanac, which had been a fairly unknown publication up until that point, had a typo that erroneously predicted snow in New England in June. And it snowed in New England in June! And from that point onward, the Farmer's Almanac became a huge best seller.

Well, it wasn't just Tambora. In fact, there were several volcanoes: St. Vincent Island in 1812; Mayon Volcano in the Philippines in 1814; and then Tambora in Indonesia. And Tambora was the largest volcano of the past thousand years. It ejected about 100 cubic kilometers of tephra into the atmosphere. All that dust and particulate matter dropped global temperatures rapidly and sent many parts of the world like Europe into a deep freeze.

Incidentally, spectacular sunsets were also observed after Tambora's eruption. Many of you may know the dramatic sunsets in the paintings of Joseph Mallory Turner. Well, his sunsets don't have those colors in the years before the Tambora eruption.

Interestingly, there was a study in the year 2007 of more than 500 paintings by many artists, such as Turner, Rembrandt, Reubens, Degas, Copley, Gainsborough, and it showed that the sunsets of these artists were significantly more colorful in years following a big volcanic eruption than before. I mean, yes, they were creative and interpretive in what they were painting, but they were also painting what they saw.

Now, I can go on. There are many examples of other significant historic volcanic eruptions that greatly changed civilization. If you go back 3,000 years ago, 1150 to 1136 years B.C.E., again, Iceland was erupting—Hekla Volcano. Ashes rained on China for ten days straight, and many anthropologists suggest that 90% of the population of Scotland and Northern England died at that time.

And 209 years B.C.E., again, probably in Iceland—an eruption caused global devastation. The Chinese historian Pan Ku said that great famines killed more than half the population. People ate each other, and the emperor lifted legal prohibitions against the sale of children. The stars weren't seen at night the following year for a period of about three months.

The year 536 C.E., Mount Rabaul in New Guinea—the Byzantine historian Procopius wrote that "the sun gave forth its light without brightness, like the moon. The darkness lasted for 18 months and each day the sun shown for only four hours." He also said that 80% of the population starved to death at that time. Population losses on the order of 80–90% are absolutely unfathomable today. But we can go back within just the last few thousand years and find several events, several times, when volcanic eruptions caused this to happen.

Incidentally, there is one eruption that may have been worse than any of these. It's interesting. If you go back and you look in the climate record, a period of severe Ice Ages began about 75,000 years ago, a little less than 75,000 years ago. If you look in anthropology at the mitochondrial DNA of *Homo sapiens*, it seems like humans living today evolved from common ancestors also about 75,000 years ago. And this is interesting, because *Homo sapiens* had evolved, we think, somewhere around 200,000 years ago and had spread into different regions of the world, had come out of Africa.

But we're not related to those people. We're related to a small group of people that seem to have all existed about 75,000 years ago. Well, why? Is there any correlation between these two? Well, 75,000 years ago, Mount Toba erupted—Toba Volcano in Indonesia. It was the largest eruption in the past 100,000 years, and 280,000 cubic kilometers of rock was ejected into the atmosphere. The explosion of this volcano was equivalent to a gigaton of TNT. If you go there now, you see the caldera from this volcano is 100 kilometers in length; it's the largest in the world.

And what some anthropologists propose is that though humans had scattered about the globe, the strain on the environment was so severe that most humans may have died. It was just in a few places, like the warm, sheltered rift valleys of Africa, where a few tribes survived, and all modern humans have since evolved from that group that survived.

In other words, a volcanic winter would have followed this eruption. But following that volcanic winter would have been an extended period of cold, probably lasting thousands of years. So, over weeks, to months, to years, to decades, to millennia, the population would have continued to drop, and dwindle, and it passed through a narrow bottleneck. That bottleneck, the tightest point there, represents the population from which all modern humans evolved. I mean, these sorts of phenomena, eruptions like Toba, don't happen very often. But they have happened in the past and, of course, they always have the potential of happening again.

Well, from this lecture, I hope that you now have a better understanding of how climate changes happen sometimes rapidly and sometimes over very short timescales. So, over the past few lectures we've examined the mechanisms of climate change over very long, intermediate, and now short time scales, and seen the ways they can affect human history long before people came into the story with the Industrial Revolution and our own production of greenhouse gases.

In the next lecture, I want to talk about some specific examples of how climate and climate change have actually shaped the course of human society and civilization. I'm going to essentially run through a timeline of human history and show how the events in the history of our civilization have been punctuated by particular changes in climate from mostly these short timescale events that I've talked about today.

Lecture Forty-One
Climate Change and Human History

Scope:

The evolution of life depends upon the natural selection of individuals within particular environments. Because local and global climates are continuously changing, the kind of life that exists on Earth is also continuously changing. The mass extinctions of life that mark out the different periods of the geologic record occurred because of changing conditions on Earth, due to both internal (volcanism, plate distribution, ocean circulation) and external (meteorite impact) factors. Mammalian and human evolution was greatly shaped by changing climates. Even the course of human civilization, which began at the same time as the warm, stable climates of the current inter-glacial period, is strongly tied to small changes in global and regional climates.

Outline

I. This lecture discusses the history of climate change and human history. Human migrations, establishment of civilizations, history of wars, and even the development of countries would all have been totally different with only a slightly different set of climate controls.

    A. The past 10,000 years have been the warmest, mildest, and most stable weather conditions for which we have evidence.

    B. When the climate returns to its usual, more variable patterns, the challenges to human existence may be severe.

II. Looking back across time, we can see that relatively small changes in climate have had significant affects on the course of human history.

    A. About 100,000 years ago *Homo sapiens* were emerging as the dominant hominid species. Anthropologists have suggested there may have been an evolutionary selection for large brains during the strong Ice Ages that occurred 120,000 to 90,000 years ago. You needed to be able to think to survive through these challenging times.

B. During 60,000–40,000 years ago there began a huge diaspora of people from Africa, spreading all around the globe. Temperatures were more stable, food sources were better, and there were more stable coastline locations. Sea levels were low, so there were land bridges between many different land masses to facilitate movement.

C. The beginnings of agriculture and the first pre-civilization communities evolved about 10,000 years ago. This was the beginning of a warm and relatively stable climate, typical of the ending of a period of Ice Ages.

D. The rapid increases in temperatures that began about 16,000 years ago were probably the result of a runaway greenhouse effect. This was a dynamic feedback resulting from warm temperatures reducing ice coverage on land, causing increased infrared radiation from more exposed Earth surface.

E. Another factor in the runaway greenhouse effect may have been a rapid release of methane from ocean sediments eroded when sea levels were low.

F. Global temperatures suddenly stopped increasing about 13,000 years ago, especially in the northern hemisphere, and dropped back into another 1,200 years of extreme cold. This was probably due to a shutdown of the thermohaline circulation in the northern Atlantic Ocean.

  1. There is evidence for a large comet exploding over Canada at this time, and the rush of melted water into the Atlantic could have caused the Gulf Stream to stop flowing.

  2. However, an episode of cooling in the southern hemisphere had begun 1,000 years earlier, and its cause is not yet known.

G. It is often not appropriate to talk only about global climate change because regional weather patterns are so variable. For example, from 22,000 to 5,500 years ago the Sahara Desert went from an arid desert to a fertile savannah and back again.

H. In spite of regional effects, however, civilization began to emerge simultaneously in several different places about 6,000 years ago. There were probably two reasons for this.

1. The sea level stopped rising, which resulted in the stabilization of shoreline communities.

2. This, combined with the warmer climates caused and a large increase in plant and animal biomass that could sustain larger communities of people.

I. The rise in sea level that followed the Ice Ages permanently forced all shoreline communities tremendous distances from their initial homelands (explaining why many cultures have myths about being expelled from an Eden).

J. Between 3600 and 2800 B.C.E. was a period of climatic deterioration, with alternating droughts and floods, which gave rise to developments such as the large irrigation and drainage systems in the Indus Valley.

K. About 2900 B.C.E. extreme flooding brought about the start of the Sumerian Empire. This may have been the basis for the story of Noah's Ark, which is from a Babylonian version of a Sumerian legend called the *Epic of Gilgamesh*.

L. Between 2200 and 1200 B.C.E., weather conditions brought about a collapse of many Bronze Age civilizations with an abandonment of agriculture and shift to more nomadic lifestyles.

M. Between 800 and 500 B.C.E. the climate moved into a warmer period, leading to prosperity in the Middle East, the spread of the Celts in Britain and the start of the Roman culture along the Mediterranean.

N. The continued warming in Asia created large-scale trade between Europe and Asia, including the opening of the "silk route."

O. Between 0–100 C.E. warm, stable climates allowed the Roman Empire to thrive and expand. However, in 400 C.E., the climate went into an extended period of freezing. Starving Europeans migrated south and eventually overrode the Roman culture, contributing to its demise.

III. The cold spell that started around the year 1300 had serious consequences for the entire world.

A. The colder climates caused severe snows in winter and cold, hard rains in summer, and this led to the Great Famine of 1315–1317.

B. In China, the cooling climate caused massive floods that drowned more than 7 million people along the Yellow River in 1332. With so many people dead, no one was left to bury the bodies, which were left at the surface. This led to a boom in the rat population.

C. The rats carried the fleas that carried the bubonic plague, also known as the Black Plague.

D. Many regions were importing grain to help alleviate the famine, and ships carrying grain from China went to the Middle East and Europe, bringing the rats with it.

E. The plague spread out from all of the port cities, eventually killing 70% of the British population by the end of the century (and a quarter to two-thirds of other regions as well).

F. The plague may also have contributed to the Little Ice Age that followed. Millions of trees sprang up in now-abandoned fields, pulling carbon dioxide out of the atmosphere and cooling the climate.

G. The plague changed the social and economic structures of Europe and Asia.

    1. The Catholic Church lost tremendous support and its membership was severely decimated.

    2. Significant racial discrimination began resulting in the permanent relocation of many Jewish communities to parts of Eastern Europe.

H. In Western Europe, the loss of population led to a loss of cheap labor, bringing about the end of serfdom and beginnings of capitalism.

    1. It began an era of free-enterprise and entrepreneurship that continued with the industrial and scientific advances of the Renaissance.

    2. In Eastern Europe, where the plague did not hit as hard, the social structure remained in place and serfdom continued until the late 19th century.

I. The whole course of history was changed forever because of rains that came with the cooling that followed the ending of the "Medieval Warm Period."

Recommended Reading:

Gore, *Earth in the Balance* and *An Inconvenient Truth.*

Mithen, *After the Ice.*

Questions to Consider:

1. Do you think civilization would have been able to evolve if there had not been such a warm, stable period of climate beginning about 10,000 years ago?

2. What do you think the effects on humans will be if the climate continues to warm precipitously? What about if the climate starts to cool and go back into a glacial period?

Lecture Forty-One—Transcript
Climate Change and Human History

Welcome. In this lecture, I want to talk about the history of climate change and human history. It turns out that the human migrations of peoples, the establishment of civilizations, the history of our wars, and even the development of countries themselves, would all have been totally different with just a slightly different set of climate controls. I want you to remember, after I go through this discussion, that the past 10,000 years have been the warmest, mildest, and most stable weather conditions that we can have. When the climate returns to previous and more normal levels, the challenges for human civilization may be severe.

Let's go back in time. Let's go back 100,000 years ago, and run the clock forward, and look and see if we can find where events in human history correlate with climate events. Back 100,000 years ago, *Homo sapiens* was emerging as the dominant hominid. There were other hominids around, like *Neanderthals*. Why? Why were we favored? Why did we survive? Well, one possibility that anthropologists have suggested is there may have been a selection, an evolutionary selection, for large brains during the very strong ice ages that occurred 120,000 to 90,000 years ago. In other words, you needed to be able to think in order to survive.

If you go 60,000 to 40,000 years ago, there was a huge diaspora of people out of Africa going into Europe, Asia, and Australia, all around the globe. They moved along the coastlines largely, at a rate of around a kilometer per year. That's the rate at which people spread out. Why? What happened then? Why didn't this happen earlier? We were genetically the same. Well, this was a period of much more stable temperatures that followed these periods of very rapid cooling, like what happened after the Toba eruption that I talked about in the previous lecture. There were better sources of food. There were more stable coastlines. The shoreline wasn't moving back and forth so rapidly. Again, as I talked about previously, humans could reach Australia because the sea level was so low that there were land bridges. Humans came to North America 14,000 years ago, crossing from Siberia to Alaska. About 50,000 to 40,000 years ago, in Europe, there was a cultural explosion at the time. Jewelry, figurines, perhaps religious icons, were made. Why?

Well, there was a warming trend in Europe, and life was easier, perhaps. Maybe there was time for other things besides just scraping out a living.

The beginnings of agriculture and the first pre-civilization communities, cities like Jericho, didn't actually evolve and occur until about 10,000 years ago. Why? What happened at this time? Well, this was the start of our last interglacial, the start of that warm and relatively stable climate. Remember, when I talked about the Milankovitch cycles, the fact that there were these oscillations in Earth's ellipticity and that the interglacial periods, the times of warm climates, corresponded to the fluctuations in the shape of Earth's orbit, that roughly 100,000 year cycle. At the end of each period of ice ages, there came a rapid increase in temperature leading to a fairly brief period of interglacial, warm temperatures.

But why were there such rapid increases 16,000 years ago? Well, the answer is probably that there was a runaway greenhouse effect. In other words, there is a dynamic feedback, here. You have reduced ice coverage on land, you now have more of Earth's surface, which emits more infrared radiation, which warms the atmosphere, which can hold more greenhouse gases, which continues to warm the surface.... You go into this cycle of runaway greenhouse effect. So, that's probably why the ice ages came to a fairly rapid end and do so at the end of each period of ice age.

There's another possibility, however, and that is the rapid release of methane from ocean sediments. This is a really interesting situation. At the end of an ice age you have really low sea levels, so a lot of former marine environment, marine sediments, are exposed above land, and they are rapidly eroding. It's interesting: If you go to the Mediterranean, you see evidence of lots of huge landslides at the end of the last ice age, of rushing off coastal regions down into the Mediterranean Sea. Well, these areas had a lot of methane in frozen offshore hydrates—I'm going to talk more about this in Lecture Forty-Three—and the sudden erosion of these sediments would have released enormous amounts of methane.

That sudden release of methane could have triggered the runaway greenhouse effect. It could have been the event that essentially kicked off that sudden global warming. Of course, there may be other causes, as well. Again, we're just beginning to figure this out.

Global temperatures begin to increase dramatically about 16,000 years ago, but then stopped 13,000 years ago, and dropped back into another 1,200 years of freezing cold. This is a period known as the Younger Dryas, named after an alpine tundra flower, the dryas. The effect was most significant around the North Atlantic. In fact, temperatures in places like Greenland dropped by 10°C. Why? What plunged us back into another short, brief ice age? Well, this was probably a shutdown of the whole thermohaline circulation of the Atlantic Ocean that I talked about in the last lecture. The Gulf Stream stopped flowing. This time there was a huge glacial lake, the glacial lake Agassiz, that released huge amounts of water off the surface of Canada into the North Atlantic. The fresh water was too buoyant, it couldn't sink, and the whole Gulf Stream stopped and shut down, sending Greenland, Europe, and North America into a period of freezing.

Why did it happen at this particular time? And why is this event not seen at the start of other, previous interglacials? Here's another unusual possible climate change event: It could be that a large comet caused this short, brief freezing. There's evidence for a large comet exploding over Canada at this time. If you look at soil, it contains the element iridium, melted carbon droplets, even tiny diamonds, the sort of materials that you would expect after some sort of meteorite or comet impact. However, we don't see any crater in Canada, so it's thought that this explosion would have occurred in the atmosphere, which would have meant it would have been an icy comet instead of a rocky meteorite. The heat from the explosion would have caused wildfires, extreme heating, and glacial melting. It would have shut off the Gulf Stream, chilling the northern hemisphere.

It's interesting. At the same time in the southern hemisphere, there had been a milder episode of cooling which had actually started 1,000 years earlier, so obviously it couldn't have been caused by this impact of a comet. It could have been caused by something else. We still haven't figured this out. Again, we have to be very careful. It's often not appropriate to talk about global climate change, because things get so heavily influenced by regional weather patterns. One of the best examples of this happened in the Sahara Desert.

From the peak of the end of the ice age, 22,000 years ago, to about 10,500 years ago, the Sahara was bone dry, even worse than today. In fact, the Sahara extended 400 kilometers farther south than today.

Humans existed only along the wet Nile River, not at all in the desert regions. But about 10,500 years ago, monsoon rains from the Indian Ocean began arriving regularly. The Sahara was transformed into a savannah environment. Wide settlements spread all across what is now Egypt, with lots of cattle grazing on grasses. About 7,300 years ago, there began to be a retreat of those monsoon rains. The prehistoric peoples had to either retreat south into what is now Sudan or move toward the Nile. By 5,500 years ago, 3,500 B.C.E., there was a full return back to the dry desert. This represented the start of dynastic Egyptian civilizations along the Nile.

All of this change was the result of regional variations in atmospheric circulation. They were affected by, though clearly distinct from, the global changes that were occurring at the same time. Now, in spite of regional effects, in general, civilizations around the world largely began about 6,000 years ago. This happened simultaneously in the Fertile Crescent around the Euphrates River, the Indus Valley in what's now India and Pakistan, the Yellow River Valley in China, along the Nile in Egypt, and along the Mississippi River Valley in North America. All of these sprung up at about the same time. Why? What was going on 6,000 years ago that allowed civilizations to suddenly start simultaneously, everywhere?

Well, there were probably two reasons. First of all, the sea level finally stabilized. If you go back 10,000 years ago, the sea levels were 50 meters lower than they are today and rapidly rising. At 9,000 years ago, the sea level was 25 meters lower than today. At 7,000 years ago, the sea level was still 7 meters lower than today. But by 6,000 years ago, the sea level was less than a meter lower than it is today. It hasn't moved very much since then, and this is very important for civilization, because it allows a stabilization of shoreline communities. You're not having to move back and forth as the sea level changes. It turns out that shoreline communities were also important for inland civilizations, because there's archaeological evidence that inland civilizations imported a lot of fish from the shorelines.

The other important factor here is simply that in a warmer climate with an elevated sea level, the plant and animal biomass increases rapidly. You have continents with flooded continental shelves, which mean lots of shallow water extending for long distances off shore.

This gives rise to large ocean shoreline fish populations. Taken together with the increase in shallow environments, and the warmer temperatures, the coastal margin of productivity, the measure of how much biomass is in the whole shoreline region, probably increased by a factor of ten. In other words, you could begin to support large communities of people.

The shoreline stopped moving. There was lots of fish and food. Inland, you had warm stable climates that could lead to good agriculture. This was the beginning of complex, large civilizations. Now, it's interesting: If you go back to the times of the earliest civilizations, many cultures have myths that are similar to the story of the expulsion from Eden. Why? Why does this story run throughout civilizations? Well, for a very simple reason: They were expelled. The rising sea levels that followed the end of the ice ages permanently forced all shoreline communities tremendous distances from their initial homelands. If you go to the Persian Gulf today, you can see archaeological remnants of previous communities that are now flooded beneath the Persian Gulf. The water flooded up in toward the Tigris and Euphrates regions, moving that fertile delta constantly inland, flooding all the communities, forcing people to leave what were once their homes.

It's interesting. Sometimes, these floods occurred catastrophically. If you look in Eastern Europe, it turns out this was the origination of several peoples who now live further south in the Middle East or even further in Africa. Some Egyptian and Semitic peoples actually originated in Eastern Europe. Why did they leave? Well, it turns out that 5,600 years before the common era, the Black Sea flooded dramatically and catastrophically. The sea level was rising throughout the Mediterranean, but that didn't affect the Black Sea initially, which was an entirely separate lake, entirely landlocked. At some point, the sea level rose high enough, and the water burst through what's now the Bosphorus, right by Istanbul. The water flooded enormous amounts—12,500 cubic kilometers of water burst through the Bosphorus and essentially created the Black Sea in a remarkably short amount of time. Whole towns are under water there. You can still see the remnants beneath the surface of the Black Sea. Those peoples all had to move elsewhere. Many cultures have similar flood myths from this time. You can go to the Baltic Sea in the north. Again, you have whole communities that once existed before the flooding of this region, but as the sea level came out, this whole area became water and people had to very rapidly leave their initial communities.

Starting about 3,600 to 2,800 before the common era, the climate went into a phase of rapid, small changes. We call that a period of climatic deterioration. It was a series of alternating droughts and floods. It's interesting, because civilizations had to change, at this point, to handle these alternating droughts and floods. The Indus Valley civilization in modern day India and Pakistan began to develop large irrigation and drainage systems about 3,100 years before the common era in order to deal with this. Larger civilizations in the Middle East region began to develop in order to survive, and this was the beginning of one of the major large empires in the Fertile Crescent, the Akkadian Empire in modern-day Iraq.

About 2,900 years before the common era, there was an extreme flooding that actually brought about the end of one empire and the start of the Sumerian Empire. Interestingly, this flooding, these extreme rains, may have been another cause of the flooding myths. In fact, they may have been the origination of the story of Noah's Ark and the biblical deluge in the Old Testament, which is actually taken from a Babylonian version of a Sumerian legend, called the *Epic of Gilgamesh.*

At the same time in Egypt, we have the story of Joseph warning the Egyptian pharaoh to prepare for seven lean years. Why was this important? Well, the climate was changing, and cultures had to be able to store enough food in order to last through these periods of drought to feed the people and survive. Those communities that were able to do that, that had the foresight to prepare, were the ones that survived and the ones that exist today. Those communities that didn't make these preparations didn't survive and were overrun by other cultures and communities.

Starting about 2,200 years before the common era, century-long droughts caused the collapse of the old kingdom of Egypt and of the Akkadian Empire in Mesopotamia. This was a very rough time. The Indus Valley culture, around 1,900 years before the common era, essentially became overrun by desert sands from the expanding desert. About 1,200 years before the common era, there was a northward shift in the winter storm paths that actually caused significant droughts in the Mediterranean and, at the same time, flooding in Central Europe. And this essentially brought about a collapse of the Bronze Age.

At this time in the Middle East, there was a large abandonment of agriculture, which just didn't work stably and was failing, and a shift by a lot of peoples to a nomadic lifestyle. There were mass migrations of peoples—the Phrygians and the Hittite peoples from what is now modern-day Turkey—southward. The influx of these people into the Middle East led to the huge numbers of battles and conflicts that are described in the Old Testament.

Things got better. Between about 800 and 500 years before the common era, we went into a period of warm and mild Atlantic climates. There was wide prosperity in the Middle East. There was the spread of the Celts in Britain, the start of the Roman culture along the Mediterranean. However, at 500 to 400 years before the common era, as I talked about in the previous lecture, the warming was so great that it caused a huge melt of Greenland ice. We think it shut down the Atlantic Ocean circulation. It temporarily froze Europe, incidentally giving rise to the Macedonian Empire, which I talked about in the last lecture.

The continued warming in Asia at this time opened up mountain passes, and it created, for the first time, large-scale trade between Europe and Asia—the opening of the "silk route." People like Marco Polo were going back and forth between Europe and Asia, allowing the two cultures to finally grow somewhat in parallel and learn from each other. Europe got a couple of things that were absolutely invaluable to its whole future, things like gunpowder and even ketchup.

If you go to about 2,000 years ago, we were just in a period of very warm stable climates, about 0 to 100 years in the common era. This allowed the Roman Empire to thrive tremendously. The Roman Empire, at that point, had more than 60 million people in it. Rome itself, the city, had about a million people and controlled lands that extended all the way from England, to the Middle East, to northern Africa.

So, why did Rome collapse? Well, again, if you've studied European history, you can probably rattle off a whole bunch of reasons that involve things like lead in the water and a general decay of the people. Well, yes, that's true, but there also was a climate effect to this. In about 400 A.D., the climate went into an extended period of freezing, a very strong cold spell, which had followed many centuries of warm, stable climates. Europeans throughout the lands were starving. Their crops were failing, and they moved south. The Huns, the Goths, the Visigoths, the Ostrogoths—these people began moving towards the Mediterranean. Again, it was the same situation as the westward expansion of the United States. What would cause peoples just to pack up their materials and move out of their homelands? They'll do it if they don't have enough food to eat. The southward migration of these many northern peoples eventually overrode the Roman culture and was a significant reason for its final demise.

A little bit later on in history, again, we had a warming period. It's interesting, because look at Leif Erickson, later on; why were Leif Erickson and the Vikings able to sail to America? Well, from about 900 to 1300, there was an extended warm period that we call the "Medieval Warm Period." At the same time, there were severe droughts in Central America and southwestern North America, and this led to the collapse of the Mayan cultures and the Anasazi civilizations in the southwestern United States.

It's an unfortunate result of warming, but when global temperatures increase, the climates in the southwestern U.S., and in Central America, get very warm and dry, and this happened at about this time. In fact, we actually have very good records for the climate change in these regions, because we have tree rings from trees that were alive at this point. The tree rings show us that there was a period from 1118 to 1179 in the common era of extreme drought. We see this in trees on the modern-day Colorado Plateau.

It was during these times that all of these cliff dwellers—Bandolier, and Chaco Canyon, and Mesa Verde—all these communities just disappeared. The cliff dwellings were abandoned. There's no sign of war, there's no sign of strife; the cultures just seemed to disappear, and we think we know why, now. It's because the climates got so dry that they simply ran out of water. They didn't have the technology to be able to drill down underground to pull water up from the groundwater, as we have today, and they just ran out of water.

Interestingly, there was a thawing of the North Atlantic ice—as I mentioned—that allowed the Vikings to establish these settlements in Vinland in what is now Labrador, Canada. With the end of that Medieval Warm Period and the cold spell that started in about 1300, the North Atlantic froze up, and all those Viking settlements got stranded. They were no longer able to sail back and forth, and those communities died off.

The cold spell that started around 1300, however, had much worse consequences for the whole world. Remember, when you have a colder climate, you reduce the amount of humidity, the amount of water vapor that the air can hold, and that falls out as rains. So, the increase in cold, the increase in high humidity—more rains. You had severe snows, you had cold, hard rains in summer, and this led to the Great Famine of 1315 to 1317. Within a few years, a quarter of the population of Europe died off—mass starvations. I mean, did you ever wonder where stories like Hansel and Gretel could ever come from? And the rest of the Grimm stories? Abandoned children, cannibalism? These things happened during the Great Famine when people didn't have any food. And it followed a period of long prosperity, so populations were high, and the populations just could not be sustained.

Interestingly, we also know why the Black Plague struck in the 1300s, wiping out more than half of Europe by 1350. Well, the story actually begins with flooding in China. The same change into a flooding, wet, cold climate caused massive floods in China. In fact, it's estimated that in the year 1332 more than 7 million people drowned along the Yellow River in China. With so many people dead, there were no people left to bury them, and the bodies of humans and other animals were just left at the surface. That led to a boom in the population of rats, and on the rats are fleas, and it's the fleas that carry the Black Plague, the bubonic plague. So, you had a boom in the population of rats and fleas, and the plague began to spread throughout China. But here's the issue: Europe, at this time, because it was starving, was importing grain from wherever it could get it. Ships of grain were being sent from China to the Middle East and Europe.

In China, the devastation of the plague was also tremendous—more than two-thirds of the population died there—but the plague spread both by land and by sea with anyone traveling westward from China.

From every port city in the Middle East and European countries, the plague spread out as those ships filled with grain carrying the rats arrived. It eventually killed one-quarter to two-thirds of the populations, finally reaching Britain in 1349, where it killed 70% of the population by the end of the century. There are some parts of Europe that have actually never returned to the high populations that they had at the start of the 1300s. There are areas in France where the modern-day populations are less than they were at that time. In other words, Europe still hasn't fully recovered from this catastrophic event. The plague wasn't done just in that year. It returned in 1563, in 1578, in 1593, 1603, 1625, 1636, and 1665. In fact, the outbreak in 1563 in places like England was actually worse than when the Great Plague first hit in 1349 in Britain.

It has even been suggested that the plague contributed to the Little Ice Age that followed. Bizarrely, you had a situation where all these fields that had been plowed for agriculture were now abandoned. Millions of trees sprang up in these abandoned fields and rapidly pulled carbon dioxide out of the atmosphere, playing a factor in cooling climates.

The plague entirely changed the social and economic structures of Europe and Asia, with more than half of the populations gone. For one thing, there was a tremendous loss of faith and support for the Catholic Church, because it couldn't do anything to stop the plague. In fact, it suffered some of the worst losses, because often the church members were trying to help with the plague, and they got exposed to it at greater rates than other people. There was also, in Europe, the beginnings of significant racial discrimination. Hundreds of different massacres of Jewish communities occurred. In fact, there occurred a permanent relocation of many Jewish communities to parts of Eastern Europe, like Poland, where the plague didn't hit very hard. There wasn't the same social and economic and political strife occurring in places like Eastern Europe.

Interestingly, there were other significant differences between the whole future of Eastern and Western Europe that stemmed because of this event. In Western Europe, the incredible loss of population led to a loss of cheap labor. It essentially brought about an end of serfdom. It started the very beginnings of capitalism. It began an area of free enterprise and entrepreneurship that continued on with the whole industrial and scientific advances of the Renaissance. This didn't happen as much in Eastern Europe, where the plague didn't hit very hard, so social structures remained pretty much in place, and serfdom actually continued until the late 19th century. The whole course of history was changed forever because of rains that came with the cooling that came with the ending of the Medieval Warm Period.

In Lecture Forty-Five, I'm going to bring this story of climate and climate change back up to the present. But this gets a bit more complicated, because humans begin to play a dominant role in the story. Before I do this, I want to talk about a few more areas where geology has shaped human history. In the next few lectures, I'm going to talk about how the distribution and formation of natural resources, like minerals, and metals, and sources of energy, like fossil fuels and uranium, have also played a significant role in affecting human history and culture and economics.

Lecture Forty-Two
Plate Tectonics and Natural Resources

Scope:

Did you ever wonder why there is gold in California, coal in Indiana, and oil in Iraq? Geologists have spent centuries wondering this as well, but it wasn't until the history of plate motions was interpreted that the causes of the distribution of the world's geology became apparent. Humans rely heavily upon Earth's natural resources (An old saying goes "If you don't grow it, you mine it."), and it is through plate motions and interactions that most mineral and petroleum resources are formed. For example, the large oil fields in the Mid-East come from underground anticlines that have formed during the Arabia-Eurasia plate collision.

Outline

I. This lecture discusses natural resources and the economic effects of geology and their connections to the history of geologic change.

    A. Your home, work environment, and modes of transportation are all based on the mineral, metal, and fossil fuel resources we get out of the ground.

    B. Every year more than 25,000 pounds of new, non-fuel minerals must be provided for each person in the United States to make the items that each of us use every day. This comprises about $2 trillion of our annual $13.2 trillion gross domestic production.

    C. The issues involved with maintaining this supply are incredibly complex; natural resources need to be available in social, political, and economic ways, as well as geologic ways.

    D. The United States has to import 100% of many of these minerals because they do not exist within our boundaries.

II. Political conflict has always resulted from the uneven distribution of natural resources.

A. The author Jared Diamond believes this uneven distribution contributed to Europe's ability to achieve dominance over the Americas and other parts of the world.

B. The uneven distribution of resources is a direct result of the geologic roulette wheel; resources just happened to be created in some places and not others, depending upon the particular history of plate motions.

C. Many of these resources are very limited and may run out within the 21st century.

D. "Mineral reserves" are a known quantity of a substance; mineral resources are projected quantities based upon previous geologic discoveries.

E. The way we handle limited reserves of petroleum and other resources will determine the future of humanity, not only in the distant future, but in the near future as well.

F. Archaeologists often define human history in terms of the sequence of discovery of what are called the "seven metals of antiquity": gold, copper, silver, lead, tin, iron, and mercury.

III. The processes of plate tectonics—ocean rifting, the motion of the ocean seafloor across the mantle, the subduction of that material, and the subsequent volcanism—concentrate metals and minerals to levels that make them usable for us.

A. The key ability of the tectonic process to concentrate these minerals is that atoms for large elements do not sit well within the silicate mineral structures.

    1. Minerals must meet two criteria: the electrical charges of the different ions have to balance to zero, and the different sizes of the ions have to match.

    2. A large atom like gold or silver does not fit well within a mineral structure; as a result, these minerals tend to be the first to dissolve and leave a rock the first chance they get.

B. This process of concentrating useful metals and minerals occurs at mid-ocean ridges, where the materials get concentrated on top of the ocean seafloor.

C. When the ocean seafloor enters a subduction zone, the minerals are concentrated further.

1. This can happen during a future continent-continent collision, with subsequent erosion.
2. This can also happen when some of the mineral-rich material gets carried down into the subduction zone.
3. If the material flows upward with the water, further concentration occurs by hydrothermal circulation at the island arc volcanoes or continental arc volcanoes through similar processes as at the mid-ocean ridge.

D. If your country has former subduction zone rock, you likely have access to concentrated metals that you would not have in a country that hasn't had subduction as part of its geologic history.

E. Continental volcanoes can also be places of mineral concentrations because of the presence of hydrothermal circulation there.

IV. Other mineral resources form in a variety of different ways.

A. Gems such as amethyst, agate, turquoise, and malachite form through precipitation of crystals directly out of water, analogous to the way salt forms.

B. Quartz-rich crystals called pegmatites form from precipitation of silica directly out of the hot fluids in the last stages of the cooling of magma underground, and can often contain a variety of valuable minerals and metals.

C. Gems like zircon, topaz, and ruby sometimes form from crystallization within pre-existing gas bubbles of volcanic rocks.

D. Garnet, jadeite, and tourmaline form in high-pressure metamorphic environments.

E. Diamonds have to form at least 150 kilometers beneath the surface. They came up to Earth's surface many billions of years ago in conduits called "kimberlite pipes."

V. The United States and other industrialized nations face several large issues concerning mineral reserves and resources.

A. The United States is dependent upon a wide range of metals and minerals that have specific and unique applications in our technological world.

1. Many of these are either very small or are being used at very rapid rates.
2. The uneven distribution of these resources has caused industrialized nations like the United States to be tremendously dependent upon foreign imports.

B. For example, at the start of the 20th century, cars were made of only five materials: wood, rubber, steel, glass, and brass. They are now made of at least 39 different minerals in addition to those five materials.

C. Some functions of cars use very specific minerals with no known substitutes.

D. Many minerals have resources that may run out within the century.
1. Humans use 15 million metric tons each year of copper because it is the best material for conducting electricity.
2. There are likely more than 1.6 billion tons of copper geologically available, but this may last only 100 years.
3. However, more resources will be discovered and extracted as the price of copper goes up, but there are not infinite supplies.

E. The best examples of minerals that have critically limited supplies are "rare Earth metals."
1. These include europium, erbium, cerium, neodymium, samarium, gadolinium, and several others.
2. Most of these rare Earth metals (about 75%) come from China.
3. Nations of the world must rely upon global trade to obtain these metals, and that requires countries to cooperate with each other.

F. Another critically limited resource involves the platinum group metals.
1. About 80% come from South Africa, while only 5% are found in North America.
2. These act as catalysts for a huge number of chemical reactions.

G. One thing that helps to extend the lifetime of many of these mineral and metal resources is recycling.

1. Recycling and reusing is a great benefit that mineral and metal resources provide over nonrenewable resources like fossil fuels.
2. As recycling becomes more economically advantageous, the projected lifetimes for many of these will be greatly extended.

VI. Another large area of natural resources is in fossil fuels: hydrocarbons like coal, oil, and natural gas.

   A. Coal forms primarily from land-based plant matter. It is sedimentary rock that is essentially fossilized swamp and bog.

   B. If shorelines are advancing and retreating, swamp material becomes deposited over broad regions like shoreline sediments. Areas like the Florida Everglades, for example, are sites of future coal reserves.

   C. Organic material goes through several stages of sedimentation and metamorphism to become the coal we use.

   D. Material dug up from a current-day swamp is "peat." It contains about 50% carbon. It will burn when ignited, but is a dirty source of energy.
      1. Over time, peat gets compacted and becomes lignite.
      2. After more time, lignite becomes bituminous coal (about 86% carbon).
      3. Eventually the coal becomes anthracite, which can contain 98% carbon.
      4. Most of the world's coal-based energy is in the form of bituminous coal, as anthracite is very rare.

   E. As an energy sources, coal is very abundant.
      1. Most coal today formed from swamps and bogs that existed during 300–250 million and 150–100 million years ago.
      2. Coal will continue to form naturally, but does so too slowly for significant amounts to be produced during our lifetimes, which is why we consider it to be a nonrenewable source of energy.

VII. Another major source of hydrocarbons exists in the petroleum products of oil and natural gas.

A. These hydrocarbons form in oceans from shallow marine sediments.

B. Unlike coal, petroleum needs special conditions not only to form but also to survive underground.

1. Organic sediments must be rapidly buried, compacted, and "cooked" from increased pressure and temperature to extract the hydrocarbons.

2. The heating and compression breaks down the complex molecules into a waxy substance called "kerogen."

3. The kerogen breaks down with further compression and heating into simpler hydrocarbons either in a liquid form (oil) or a gas form (natural gas).

4. Unlike coal, oil and natural gas are mobile, lighter than rock, and will eventually rise to the surface, where bacteria consume them and return the carbon into the surface carbon cycle.

C. Special geologic conditions are required in order to trap these hydrocarbons underground.

1. You need either a domelike structure, an anticline, or some other structure where the overlying rock traps these liquids and gases as they rise up.

2. You also need a reservoir material like sandstone to hold the petroleum that is both porous and permeable. Oil and natural gas can be pumped out by drilling into the sandstone.

D. Petroleum forms in several different tectonic settings.

1. A common environment for finding petroleum is an incipient rift zone.

2. When sea levels drop, water evaporates, and a layer of salt is left. When ocean water floods back in again during a period of warmer climate, a layer of organic marine sediment gets deposited on top of the salt layer. Over time you get alternate layers of salt and organic sediments.

3. With continental collisions, folding of rock underground occurs, and the sedimentary layers can form large anticlines. That is what happened in the Middle East.

4. Geologic events going on among Arabia, Africa, and Asia, creating the large underground petroleum reserves,

are just happenstance of where we are now in the plate tectonics process.

5. Ironically, the Persian Gulf was the cradle of civilization, and now the same region is once again the focal point for the world's civilizations, but for a very different reason: That's where the petroleum is.

Recommended Reading:

Diamond, *Guns, Germs, and Steel.*

Yergin, *The Prize: The Epic Quest for Oil, Money and Power.*

Questions to Consider:

1. Why are coal and petroleum called "nonrenewable" resources if they continue to be made in many parts of the earth?

2. Why is the prospect of future use of precious metals better than for fossil fuels?

Lecture Forty-Two—Transcript
Plate Tectonics and Natural Resources

Welcome. In this lecture, I want to talk about natural resources. Now, I have a theme going here about how the earth affects humans and human civilization, and there's no better example of this than in the natural resources we've been able to use from the earth. I mean, in the previous lecture, I talked about how geologic events in the form of climate change affected the course of human history. Well, here I'm going to talk about the economic effects of geology and the history of geologic change.

After all, as the old mining saying goes, "If you don't grow it, you mine it." We don't live in caves any more or wear animal skins. Why is this? It's natural resources that we've taken out of the ground. Look around you, right now. And I don't just mean at the TV. Look at your home, your work environment, your modes of transportation—all of it is based on the mineral, metal, and fossil fuel resources that we get out of the ground. I mean, even the clothes that you're wearing—if they are polyester, or rayon, or nylon, etc.— they're made from petroleum products.

I bet that you're not aware of the magnitude of this. I mean, every year more than 25,000 pounds, 11.3 metric tons, of new, nonfuel minerals must be provided for you and for each person in the United States to make the items that each of us use every day. It comprises about $2 trillion out of our annual $13.2 trillion gross domestic production. The issues involved with maintaining this supply of minerals are incredibly complex, far beyond the scope of this lecture or this course. Because for the natural resource to be available, it has to be available in many different ways, not just geologically, but technologically, politically, economically, socially, and environmentally. There are some really tricky issues involved. For instance, the United States has to import 100% of many of these minerals. They just don't exist within the boundaries of this country. We're entirely dependent upon other countries for them.

Suppose a country that has a monopoly on a particular resource doesn't want to export it? I mean, what does the rest of the world do about it? An unending source of political conflict has always come from the uneven distribution of these resources, and always will. For example, as Jared Diamond pointed out in his acclaimed book *Guns,*

Germs, and Steel, this uneven distribution significantly contributed to Europe's ability to achieve dominance over the Americas and other parts of the world. For instance, when Cortez arrived in Central America, his army of a couple hundred men defeated an Aztec army of many thousands without a single loss of life. He had horses and he had steel swords.

This uneven distribution of resources is a direct result of the geologic roulette wheel, the chance history of things like plate tectonics. Metal ore deposits, the location of oil fields, the existence of coal seams—all of these are the result of particular geologic events that just happened to create the resources in some regions and not others.

Another serious issue is that many of these resources are very limited. The reserves for many, at least, inexpensive reserves, may well run out within this century. By the way, I need to define two terms. A "mineral reserve" is a known quantity of a substance. It's what we know exists down there. "Resources" are projected quantities of these minerals based upon previous geologic discoveries. It's reserves that we know we don't know.

A lot of discussion has occurred about the limited future of petroleum reserves, but this turns out to be true for minerals and metals as well. And how we handle this issue is really going to determine the future of humanity, not only in the distant future, but in the near future as well. I mean, the discovery of metals controlled our very history—the Stone Age, the Copper Age, the Bronze Age, the Iron Age. In fact, archeologists often define human history in terms of the sequence of the discovery of what are called the "seven metals of antiquity." Gold (and how it could be used) was discovered about 8,000 years ago. Copper—6,200 years ago. Silver—6,000 years ago. Lead—5,500 years ago. Tin—3,750 years ago. Iron—about 3,500 thousand years ago. And mercury—a little less than 3,000 years ago. Each of these metals has very different properties that could be used for a variety of different mechanisms.

Let me get back to geology for a moment. Fortunately for us, naturally occurring geologic processes concentrate metals and minerals we use in everyday life to levels that make them useable for us. It allows us to take them out of the ground with a real minimum of effort, and this is due to the way that plate tectonics works. It's just the natural sequence of ocean rifting and the motion of the ocean seafloor across the mantle, the subduction of that material and the subsequent volcanism.

For instance, the most valuable metals and minerals exist within Earth's crust at incredibly low concentrations. It would really be impossible for us to use them if they weren't naturally concentrated. For example, gold consist of about three parts per billion of Earth's crust. That's 0.0000003% of Earth's crust. I mean, in order to make a wedding band, you would need to grind up and extract the gold from 3,000 tons of crustal rock. That's just not going to happen. Fortunately, geologic processes will do this for us. The process of plate tectonics concentrates metals and these minerals in ways that prevent us from having to do it. Now, there are exceptions. Things like sand and gravel exist at concentrations that are high enough for us to use right out of the ground, but for most minerals and metals, we need these tectonic processes to concentrate them.

What's the key to this? The key is—it's something I've talked about earlier—that atoms for large elements don't sit well within the silicate mineral structures. Remember when we were forming minerals I said the minerals had to meet two criteria: the electrical charges of the different ions had to balance to zero, but the different sizes of the ions also had to match. Well, if you have a large atom like gold, or silver, or tungsten, it just doesn't fit well within that mineral structure. As a result, these minerals tend to be the first to dissolve and leave a rock at the first chance they get, and this process occurs for us at mid-ocean ridges.

Remember that I talked about the process of having hydrothermal circulation through the mid-ocean ridge system, where you have heat there that causes water to expand and come out of the seafloor in these large smokers, these large thermal chimneys. In the process, it pulls water down into the rock for a wide region all surrounding the ridge. As that water enters the rock, it begins to heat up, and as it begins to heat up, it begins to dissolve that material right out of the rock. So it focuses, it concentrates, already, materials like gold, silver, and tungsten. All these heavy metals get pulled into the hot liquid. As the liquid comes out of the mid-ocean ridge, it gets instantly chilled. It solidifies out, and that's why you get this black water coming out of these thermal vents. That deposits these mineral reserves either in the chimneys themselves or the surrounding region. That material then moves away from the ridge over tens to hundreds of millions of years, gradually buried by successive ocean sediments in the process. So, already, we've had a factory at the mid-ocean ridge for pulling out valuable elements and laying them on top of the ocean seafloor.

Well, the story doesn't end there, because the ocean seafloor enters a subduction zone, and that concentrates the minerals even further. Some of the sediments get scraped off and get formed into these large accretionary wedges on the edges of continents. At some later point, perhaps in a continent-continent collision, that accretionary wedge becomes part of a mountain range eventually exposed with erosion. And the process of erosion can further concentrate them. For instance, if you go to places like California, people will pan for gold. Well, that's because the gold veins there are exposed and eroded, get carried down in streams, and people can sift it out of the streams, pan for gold.

Some of the material gets carried down into the subduction zone itself. Here it can flow upward with the water. It can flow into the melts, and then it can crystallize beneath these arc volcanoes. Then a tremendous amount of further concentration occurs by hydrothermal circulation at the island arc volcanoes or continental arc volcanoes themselves. In other words, you have a similar process going on at these volcanoes that you had at the volcanoes at the mid-ocean ridge. You have water pulled in, heated, coming out, dissolving these rich metal atoms, concentrating them, and then putting them in very narrow, particular regions.

So, it's not a coincidence that gold rushes have occurred near current or former subduction zones, places like California and Alaska. Subduction zones tend to have the richest metal and ore deposits. So, if your country has former subduction zone rock, you have access to these concentrated metals that you don't have in a country that hasn't had subduction as part of its geologic history.

Continental volcanoes can also be places of mineral concentrations. Hydrothermal circulation, again, can pull that material out of the middle of a continent if you have volcanoes forming in that particular region. Over time, erosion will expose these bodies of ores that have formed, allowing humans to extract them.

Now, there are other resources as well that form in a variety of different ways. For instance, diamonds and other gems—how do they form? Well, it's interesting. They form from other processes. For instance, you can get precipitation of crystals directly out of water, similar to how salt forms. Silica-rich fluids become minerals like amethyst and agate. Copper rich fluids become minerals like turquoise, malachite. You can also get the precipitation of minerals directly out of hot fluids in the very last stages of the cooling of a body of magma underground. They can form cracks of quartz-rich crystals called pegmatites, and these pegmatites are often locations of lots of other minerals and metals.

Some different gems require the presence of very particular elements to form. For instance, the presence of beryllium can lead to the formation of gems like beryl, emerald, [and] aquamarine. The presence of boron can lead to the formation of gems like tourmaline. You can also get crystallization within preexisting gas bubbles of volcanic rocks. Interestingly, gems like zircon, and topaz, and ruby sometimes form this way. You can also get gems forming in high pressure metamorphic environments. Garnet, jadeite, and tourmaline will begin to form in these ways.

Of course, we have the interesting case of diamonds, which, as I mentioned previously, don't form anywhere near the surface. They have to form at least 150 kilometers beneath the surface. How do those diamonds get to the surface? Well, it's a really interesting process that we really don't understand. But it turns out that early on in Earth's history, in the oldest archaic cratons, the cores of continents, there occurred rapid, explosive, high-pressure ejections of rock from way deep within the earth, from hundreds of kilometers. We find them now at the surface, that's what we call "kimberlite pipes." The kimberlite is named after the location—Kimberly, South Africa. Many of these have been found in some of the oldest cratonic rock in South Africa.

These must have been incredibly violent explosions, because these kimberlite pipes involve tremendous amounts of rock, all jumbled together, that we think some of it actually got ejected up and fell back down into the pipe. They are rare, and more importantly, they haven't happened for billions of years, but they are the locations where diamonds came up to the surface.

The United States and other industrialized nations face several large issues concerning mineral reserves and resources. First of all, we're incredibly dependent upon a wide range of metals and minerals that have very specific and unique applications in our modern technological world. Many of these resources are either very, very small, or are being used, pulled out of the ground, at very rapid rates. In addition, the uneven distribution of these resources causes industrialized nations like the United States to be tremendously dependent upon foreign imports. Let me give you an example. At the start of the 20th century, cars were made of essentially five things: wood, rubber, steel, glass, and brass. Well, nowadays modern cars are made of at least 39 different minerals in addition to rubber, plastic, and organic materials, et cetera.

Some functions of the cars, like catalytic converters and the electronics, use very specific minerals with no known substitutes. For instance, the catalytic converter in a car, especially for diesel engines, requires platinum. We don't know of any other element that will replace this function. For instance, the element indium is used for liquid crystal displays in cell phones. We don't know of any substitute, yet, for this.

As I mentioned before, many minerals also have reserves that may run out within the century. Gold, copper, silver, and tin—we're pulling these out of the ground at incredibly fast rates, and we may simply use them up. Humans use 15 million metric tons each year of copper alone. It's the best material for electricity. We don't have a good replacement for it. I got a sense of how important copper is when I visited the biggest hole in the world. It's the Bingham Copper Mine in Salt Lake City. It's a mine that's two and a half miles wide and half a mile deep. It's just a tremendous amount of ground that has been dug out in order to access this copper. Now, there are likely more than 1.6 billion tons of copper geologically available. But, at the rate of 15 million metric tons each year, that's 100 years. Of course, more resources will be discovered, and exploited, and dug out as the price goes up. But nonetheless, these are not infinite supplies.

Perhaps no better example exists for this than in what are called the "rare Earth metals." Many of these metals have unique applications in industry, and, as I mentioned, do not have any substitute. For instance, the rare earth metal europium is used for the red phosphor in color TV's and LCD screens. Even though prices are now more than $2,000 for a kilogram of this, that's 2.2 pounds, we don't know of any substitute. It's the only thing that gives us that red color.

The metal erbium is used in all fiber optic cables because of its unique optical properties. Again, it has no substitute. The metal cerium is used to polish almost all our mirrors and lenses because of very unique chemical and physical properties. Rare earth elements like neodymium, samarium, gadolinium, dysprosium, and praseodymium are used for high performance, permanent magnets in our electronics, our video games, military devices, disc drives, DVDs. There are no substitutes for them. These are materials you never hear about in everyday life, but you are absolutely dependent upon them. If you ever use a cell phone, you need these. There are other uses for yttrium, and lanthanum, and terbium, and all these other materials, and we import 100% of these. We have almost none left in the United States. Most of these rare Earth metals—three-fourths—come from China. Now, the good thing is that there is still centuries' worth in terms of the amount of these elements left. But, the whole world must rely upon global trade to get at them, and that just requires countries being nice to each other.

Another critical resource is platinum group metals like platinum, rhodium, osmium, and iridium. They're also very rare, and 80% come from South Africa. About 5% exists within the borders of North America. These metals have a unique property: They act as catalysts for a huge number of chemical reactions. As I mentioned, platinum is the only thing that works for diesel catalytic converters. Rhodium is used for removing nitrous oxide emissions. It doesn't have a known substitute.

Now, there is one thing that is already helping tremendously to extend the lifetime of many of these mineral and metal resources—recycling. It's a real advantage of mineral and metal resources over nonrenewable fossil fuels, for example, because in many cases, you can continuously reuse metals. As recycling becomes more and more economically advantageous, it is greatly extending the projected lifetimes for many of these resources.

Now, there is a whole other large area of natural resources that I haven't talked about yet, and it's what most people think of when they hear the term natural resources. That's hydrocarbons—coal, oil, and natural gas. I'm going to talk more about them in the next lecture as well, but what I'm going to do is give you a sense of just how these things form and how we get at them.

Coal, like diamond, also forms from carbon, but in a very different way. Coal is a sedimentary rock that's essentially fossilized swamp and bog. It requires a particular condition where organic material gets buried in an anaerobic condition—in other words, in an environment without a lot of free oxygen. Why is this important? Because it doesn't get eaten, primarily by bacteria, before it's buried. Most organic material—when plants die, bacteria and other organisms eat them, and the carbon gets brought right back into the surface carbon cycle. But if that material sinks into the bottom of a swamp or a bog without a lot of free oxygen, it doesn't get eaten, and it gets buried and essentially gets removed from the surface carbon cycle.

Now, if you have shorelines advancing and retreating, you get swamp material deposited over broad regions like shoreline sediments. Go to Florida now—the Everglades. These areas are future coal reserves. As these regions get buried, especially with a rising sea level, all this swamp material's going to get buried over. And, in a few hundred million years, it will be coal, it will form what we call "coal seams," layers of coal that exist, extending horizontally like other sedimentary rocks.

Now, the organic material that gets buried goes through several stages of sedimentation and even metamorphism to get the coal that we use. The point is that you've got to compact it, you've got to squeeze out other stuff, and you've got to concentrate the carbon.

Now, if you go to a current-day swamp, dig up the stuff and dry it, you've got what's called "peat." You go to many parts of the world, and people still heat their homes from dried peat. It's about 50% carbon. It will burn, but it's a very dirty, smoky source of energy, because it's 50% other stuff as well.

If it gets compacted over time, first it becomes a coal called lignite, and then bituminous coal, and eventually, anthracite. For instance, in bituminous coal, the material is about 86% carbon, and in anthracite, you get up to 98% carbon. Essentially everything else has been removed and squeezed out. Anthracite is obviously the most desirable, But it's actually very rare. Most of the world's energy in the form of coal comes from bituminous coal.

Because coal requires certain geologic conditions to form—you need a swamp—it's not found everywhere, But because swamps and bogs have been in a lot of different places, coal is actually fairly abundant as far as resources go. Most of the coal that we get comes from swamps and bogs that existed between two periods of time: between 300 to 250 million years ago and about 150 to 50 million years ago. So, the coal that we use is really old. New coal is going to form continuously, but it forms too slowly for much to occur at all during our lifetimes, and that's why we consider this a nonrenewable source of energy.

Now, this coal formed back during warmer climates; the sea level was high, continental shelves were flooded with shallow waters. The material was buried deeply in basins, particularly if you were next to eroding mountains, and that gave us huge supplies of bituminous coal. In particular situations where you had metamorphism occurring—let's say, on a continental collision—that bituminous coal can be converted into anthracite.

The other major source of hydrocarbons exists in petroleum, oil, and natural gas. These also come from carbon based organic materials, but in an entirely different setting. Petroleum, oil, and natural gas form from shallow marine sediments, mostly zooplankton and algae. And unlike coal, petroleum needs special conditions not only to form, but to survive underground.

First of all, you need to rapidly bury and compact the organic sediments, and with increasing pressure and temperature, you essentially cook these sediments, and you begin to extract the hydrocarbons out. That process of heating and pressing the material breaks down the complex organic molecules into simpler, waxy hydrocarbon molecules called "kerogen." This is the stuff that's found in tar shales. With further compression and heating, the kerogen breaks down and forms simpler hydrocarbons, either in a liquid form (oil) or in a gas form (natural gas). However, unlike coal, oil and natural gas are mobile. They are lighter than rock, so they are going to rise to the surface. Almost all the petroleum that has happened in Earth's history has quickly seeped right back up to the surface, entered back into the carbon cycle, been consumed by bacteria, and removed from the ground. It's interesting. There's a town in western Pennsylvania called Slippery Rock. Where did this name come from? It's a place where these hydrocarbons are currently seeping right to the surface, and the rocks in the streams have a slippery feel because of the presence of these hydrocarbons in the water.

Special geologic conditions are also required in order to trap these hydrocarbons underground, therefore, and keep them from rising up. I talked about this previously when I talked about the folding of sedimentary layers. You need either a domelike structure or an anticline, where the sedimentary rock forms a fold that has a peak in the middle where these liquids and gases, as they rise up, can be trapped. You need to have some sort of a reservoir material that's usually both porous and permeable. Sandstone is usually the ideal rock, and it needs to be overlaid by some cap rock, an impermeable layer, usually shale. So, as the hydrocarbons rise up, they accumulate in the sandstone. They are covered over by the cap rock. And if you drill through the shale into the sandstone, you can quickly pull out the oil and natural gas.

Now, petroleum has formed in several tectonic settings. The most common is actually in rift zones. Now, I already talked about rift zones in terms of the formation of salt layers. Remember how you can have a situation where the ocean can flood in, a growing ocean, like the Red Sea, that's just beginning to rift apart, but if the sea level drops, the water evaporates, and then you end up with a layer of salt. Well, when the ocean floods back in again at a period of warmer climate, of higher sea levels, you get rich marine life on top of that salt layer, which then can dry out again. Over time, you get alternate levels of salt and these organic sediments. That's exactly what happened with the Gulf of Mexico. There are rich, organic sediments that formed when the Gulf of Mexico was forming, when Africa was rifting away at the breakup of Pangaea. These sediments are trapped, often by the salt layers that exist on top of things.

With continental collisions, you can get the folding of rock underground, and the sedimentary layers can form large anticlines. That's what happened in the Middle East. Arabia collided with Asia during the close-up of the Tethys Sea, and two-thirds of the world's petroleum lies in a few anticlines beneath the Persian Gulf area.

In other words, it's just the happenstance of where we are now in our plate tectonic process, of what's going on between Arabia, and Africa, and Asia, that has allowed that petroleum to not only occur at a period of rifting, but then to accumulate under these very broad anticlines. I mean, who knew hundreds of years ago that beneath the desert sands of Arabia would lay an ocean of oil?

I mean, it's ironic, right? The Persian Gulf was the cradle of civilization, which I talked about in the previous lectures—Mesopotamia, Acadia, Sumeria, and Babylonia. And now that same region is the focal point for the world's civilizations again, but for a very different reason; that's where the petroleum is. In the next lecture, I'm going to continue to talk about energy resources that involve fossil fuels, and I'll talk about some of the issues that surround the use of coal, oil, and natural gas.

Lecture Forty-Three
Nonrenewable Energy Sources

Scope:

Humans consume energy at an incredible rate—more than 15 trillion watts. Most of human industry runs on oil, natural gas, and coal, which are the fossil remains of ancient dead organisms. However, the process of turning marine organic sediments or ancient forests and swamps into petroleum and coal takes an incredibly long time, and most of these reserves are hundreds of millions of years old. As a result, we have a limited supply of them. Oil and natural gas form when the remains of ocean organisms get deeply buried by subsequent sedimentation. Methane gas hydrates, layers of methane-bearing ice found in off-shore sediments, provide another source of hydrocarbons. Nuclear fission provides another kind of energy source altogether, but it is also nonrenewable because Earth has limited supplies of uranium and plutonium, which get destroyed in the fission process.

Outline

I. Vitally important natural resources that are extracted from the ground include the nonrenewable sources of fossil fuels and nuclear power.

   A. Few topics are as important, relevant, all-encompassing, and politically charged as energy. Wars are fought over it, the future depends upon it, and most of it comes out of the ground.

   B. Humans are consuming energy at a rate of about 15 terawatts (TW), which is one-third of the rate at which Earth is cooling off and losing its heat out into space.

   C. Eighty-six percent of the energy we consume comes from carbon-based fossil fuels. Including nuclear fission, 92% of all our energy comes from nonrenewable sources—sources created by geologic processes that are far too slow to be significantly produced during our lifetime.

   D. More than one-third of the world's energy supply comes from oil.

1. Twenty-five percent comes from coal.
2. Twenty-three percent comes from natural gas.
3. Six percent comes from nuclear fission.
4. Five percent comes from hydroelectric power on rivers.
5. Three percent comes from everything else.

E. Production of electricity is not efficient; we are only able to use about 40% of the energy that goes into making electricity.

II. Almost all of our energy comes from fossil fuels. This is not really a feasible long-term plan because they exist in limited resources.

A. Unlike mineral and metal resources, nonrenewable fuels cannot be recycled and reused.

B. Energy use and energy sources are not evenly distributed around the world.
   1. Industrial nations have a very small portion of the world's population but use most of its energy.
   2. For example, India uses about one-twentieth of the energy we use in the United States, but has one-sixth of the world's population.
   3. China's population is four times that of the United States and consumes about as much energy now as America does.

III. Because of the important role oil plays in powering the whole industrial complex of the world, it is arguably the most important substance in the world.

A. Oil is consumed at a rate of about 31 billion barrels every day. World reserves add up to about one trillion barrels, which means we have about 32 years worth of oil reserves.

B. The amount of new reserves discovered each year, however, roughly equals the amount used.

C. The United States uses about one-fourth of the world's oil but only has about 2% of the world's reserves.
   1. The U.S. oil supply would last less than 3 years without the addition of foreign oil.
   2. The world is in a politically unstable situation regarding oil because its dependence upon just a few countries.

D. The distribution of petroleum throughout the world is very uneven.

    1. Five countries—Saudi Arabia, Iraq, the United Arab Emirates, Kuwait, and Iran—have two-thirds of the world's oil.

    2. Saudi Arabia covers just one-third of 1.0% of the earth's surface and contains just 0.4% of the world's population, but owns 25% of the world's oil.

E. Using seismic imaging, all parts of the earth's crust have been well-examined, and we can identify where rock layers are saturated with liquid and gas. We are not missing any big oil fields.

F. Most petroleum is in liquid form, but other sources include tar sands and shales that contain raw kerogen. This is a much more expensive source of petroleum because it is harder to extract.

G. The ease of taking liquid oil out of the ground and then converting it into gasoline makes it the fuel of choice when mobility is important. Most of the world's transportation uses liquid petroleum.

IV. Another major source of hydrocarbons is that of natural gas.

    A. Natural gas is primarily methane, which has a formula of CH_4 (one atom of carbon to four atoms of hydrogen). Natural gas also contains other hydrocarbon gases like ethane, butane, and propane.

    B. About 7% of the natural gas pulled from the ground contains helium.

        1. Helium is a noble gas—it does not bond with anything else.

        2. Helium is one of the by-products of the radioactive decay of uranium and thorium that heats Earth's mantle and drives the thermal convection that drives plate tectonics.

        3. Most helium rises to the surface and flies off into space, but some gets trapped beneath the same underground capstones that trap oil and natural gas.

    C. Because natural gas is a gas, it is easily piped, compressed, and even liquefied.

D. World reserves for natural gas are about 180 trillion cubic meters.

   1. We use about 2.6 trillion cubic meters of natural gas per year.

   2. Our known natural gas reserve should last about 70 years.

   3. More is found all the time, but we do not know how much longer our resources will last.

   4. Most of the world's natural gas reserves are in Russia, Iran, and Qatar.

   5. The United States has about 5.7 trillion cubic meters of natural gas and uses about 600 billion cubic meters per year. Without imports, the United States has about 9 years' worth of natural gas.

V. The other major source of hydrocarbon is coal, which is fossilized swamp matter.

   A. Unlike oil, coal is more abundant and much more evenly distributed.

   B. The known world coal reserves are about one trillion tons. The world consumes coal at the rate of about 5.5 billion tons per year, so the known reserves would last about 180 years.

   C. Coal resources are probably much larger and will probably extend the lifetime of coal to many centuries.

   D. The United States has more than one-fourth of the global coal reserves. At its rate of use, its known supply would last about 275 years.

   E. Coal is well-distributed throughout the world.

   F. China is the largest consumer of coal, using 1.3 billion tons per year (a rate that is rapidly increasing).

   G. Because coal is solid, it can only be removed by mining.

      1. Surface strip mining is now replacing underground mining.

      2. Strip mining is cheaper, but has serious consequences for the surface.

      3. U.S. strip mining is largely done with a process that involves immediately reclaiming the land.

H. A major concern with coal is that burning it adds a lot of carbon dioxide to the atmosphere.

    1. Of the 7 tons of carbon we put into the atmosphere each year, most of it is from coal; this seriously contributes to global heating.

    2. Burning coal also adds a variety of pollutants like sulfur oxides and nitrous oxides into the atmosphere.

I. Current technologies can remove more than 95% of the carbon from coal. In the long term, it is cheaper to remove the carbon from the exhaust than to manage the consequences of not doing so; as technologies improve, the process will also get cheaper.

J. We should not be pessimistic about the problems of energy sources. Humans have become the greatest agent of geologic change, but we have the ability to handle problems as long as we understand what the problems are and make that understanding widely known.

VI. One more source of hydrocarbons that is very exciting and very interesting is that of "methane gas hydrates," also known as "methane clathrates."

A. These are frozen deposits of methane-containing ice. It looks like snow, but if you light it, it will burn. As the ice evaporates, methane trapped within the ice begins to be released.

B. Clathrates form naturally in marine sediments, permafrost, and tundra regions as a by-product of bacterial processes or from a thermal breakdown of more deeply deposited organic sediments.

C. There are enormous quantities of methane clathrates.

D. There is a large engineering challenge in getting clathrates out of the ground without having them evaporate and be lost into the atmosphere.

E. Clathrates have been discovered in many regions all around the world.

F. The amount of carbon stored in clathrates is very high. Recent assessments have brought initial estimates down somewhat, but it is still likely that clathrates contain about half as much carbon as all other fossil fuels and 10 times as much carbon as natural gas resources.

G. There is a growing sense that methane gas hydrates play an important part in rapid climate change.

1. The rapid heating at the end of each of the recent Ice Ages was probably due to a runaway greenhouse effect from the rapid release of frozen methane hydrates.

2. We need to keep close track of these clathrates because they have the potential to quickly release a tremendous amount of methane.

VII. The last major area of nonrenewable energy is in the form of nuclear fission.

A. Nuclear power is not a fossil fuel, but it is a nonrenewable energy source because we have limited amounts of uranium and plutonium.

B. Nuclear fission involves the controlled release of energy from splitting either uranium-235 or plutonium-239 isotopes.

C. About 15% of the world's electricity is generated from nuclear fission.

1. About 21% of the energy in the United States comes from nuclear power generated by 103 nuclear reactors at 64 different power plants.

2. Some countries in Europe get more than 50% of their electricity from nuclear power.

D. During the nuclear fission process, a neutron bombards a uranium isotope at high energy and breaks it apart, forming two smaller atoms (krypton and barium) and two highly charged neutrons that fly off and continue a chain reaction. A small amount of mass gets destroyed in the process, converted into radiation energy.

E. Usually, in a nuclear reaction, uranium is stored in rods which are submerged in water. The heat gets absorbed by the water, and the chain reaction is controlled.

F. The radiated energy is used to turn water into steam and drive turbines to make electricity. It is enormously efficient; 1 kilogram of uranium produces as much energy as 1,500 tons of coal.

G. The world reserves of uranium oxide are currently 3.5 million tons. The United States' resources are roughly 10 million tons.
   1. The world consumption rate is about 75,000 tons of uranium ore per year.
   2. This would provide 100 to 150 years of nuclear power at the current usage.
H. The large upside of nuclear fission is that it does not produce any greenhouse gases.
I. People have made serious objections to nuclear power on both political and environmental bases.
   1. Reactor meltdown has always been a large concern; there are additional risks of a reactor meltdown in the context of global terrorism.
   2. A more important and larger concern is the disposal of radioactive by-products. During the past 20 years, the United States has created an estimated 28 million tons of radioactive waste with no centralized location to put it.
   3. There is a lot of debate and discussion as to where to put these wastes.
   4. Transportation of radioactive waste across the country is another safety issue.

VIII. In summary, there are two major forms of nonrenewable energy sources: fossil fuels and uranium.
   A. They power the major part of our world's economy.
   B. They are going to be around for our lifetimes and our grandchildren's lifetimes.
   C. Over time, however, the increased problems of dealing with by-products and dwindling resources will make them less attractive, and the world will shift toward renewable energy sources.

Recommended Reading:

Aubrecht, *Energy.*

Deffeys, *Beyond Oil: The View from Hubbert's Peak.*

Questions to Consider:

1. It has been estimated that the true costs of gasoline are more than double what we pay at the gas pump. What do you think the major contributors to this are?

2. Imagine that you are an engineer tasked with getting methane clathrates up from 1000 meters deep in shallow offshore marine sediments without their melting. How would you do it?

Lecture Forty-Three—Transcript
Nonrenewable Energy Sources

Welcome. In the next two lectures, I'm going to talk about energy sources. In this lecture, I'm going to talk about nonrenewable sources of fossil fuels and nuclear power. And in the next lecture, I'm going to talk about renewable sources, largely driven by the sun.

Few topics are as important, relevant, all-encompassing, and politically charged as energy. It's just an incredibly important resource for all of humanity. Wars are fought over it, the future depends upon it, and currently, most of it comes out of the ground.

A big part of the issue is that humans just have an incredible need for energy. We consume it at a remarkable rate of about 15 terawatts. Remember that I defined a watt as being a rate of energy in joules per second? Remember that I defined one joule as the amount of energy to lift a plum 1 meter, so 15 terawatts is 15 trillion plums lifted up each second. Also remember that, earlier in the course, I talked about how the whole rate at which Earth is cooling off is about 44 terawatts. So, we are consuming energy almost at the rate that the earth is actually cooling off and losing its heat out into space. That's an enormous rate, and it's continuously increasing.

Of the energy that we consume, about 86% currently comes from carbon-based fossil fuels. If you include nuclear fission, the splitting of uranium, 92% of all our energy comes from nonrenewable sources. When I say nonrenewable, what I mean is that they are created by geologic processes that are far too slow for any appreciable amounts of these materials to be created during our lifetimes. Obviously, they form naturally, over the course of Earth's history, but Earth is really old, and it takes a long time to make this stuff.

Interestingly, both fossil fuels and nuclear power have several things in common. First of all, they are both mined. They both come out of the ground. They are nonrenewable. There are big questions about how much of them is in the ground still. There are concerns on how to get it out, and it turns out that there are even bigger questions on how to put stuff back into the ground. In the case of nuclear power, it's the radioactive waste that's left over from the reaction process of splitting apart the uranium. For fossil fuels, it's how to take carbon out of the exhaust from burning them and putting that carbon back into the ground.

Now the world's energy supply is provided not only by fossil fuels and nuclear power, but also by other sources as well, but when I break down these sources, you'll see how much the whole energy budget is dominated by these fossil fuels. Thirty-eight percent of the world's energy, consumed in all different purposes, currently comes from oil. That's more than one-third. Twenty-five percent, one-quarter, comes from coal; 23%, essentially another one-quarter, comes from natural gas; 6% from nuclear fission; 5% from hydroelectric power on rivers; and 3% from everything else.

Now, one-third of all the world's energy is about 5 terawatts. Five trillion joules per second goes into making electricity. Though, it's interesting: we actually only get about 2 terawatts of that power. Why? Because the production of electricity is not terribly efficient. There is so much lost in the transfer of a source into electricity and in the transmission of electricity through power wires that we only get about 40% of the actual energy that goes into making electricity.

Though almost all of our energy comes from fossil fuels, as I think you'll see after these next two lectures, it's really not a feasible long-term plan because we have limited resources. The questions are, How long do we have? How do we make the transition to something else? And what is that something else that we will eventually transition to to supply our energy needs?

Now, one important issue that we have to deal with when we're talking about fossil fuels and nuclear power is that, unlike mineral and metal resources, nonrenewable fuels can't be recycled and reused. You use them once, and they're gone. Another very important factor is that energy use and energy sources are not evenly distributed around the world. The industrial nations have a very small portion of the world's population, but use most of the energy.

For instance, the United States uses energy at a rate of about 11.4 kilowatts per person. That's equivalent to 285 40-watt light bulbs, running constantly, for each person. In fact, there's only one other country in the world that uses energy at a higher rate per person, and that's Canada, because they have such large heating needs during the winter. In contrast, if you look at developing countries like India,

India uses power at a rate of about half a kilowatt per person—one-twentieth what we do in the United States.

It's interesting. If you look at a country like China, China is actually going to play a very important part in the whole energy equation, because it's actively transitioning from a developing country to an industrialized country. It now uses energy at a rate of about 2 kilowatts per person, and because its population is four times that of the United States, it now consumes about as much energy as America does.

Because of the important role that oil plays in powering the whole industrial complex of the world, oil is arguably the most important substance in the world. It gets consumed by all different purposes at a rate of about 31 billion barrels every year. Now, how long is this going to last? Well, if you look and add up our reserves all around the world, it adds up to about a trillion barrels of oil. Do the math. A trillion barrels at about 31 billion barrels per day will give you 32 years.

However, it's not so simple. Our reserves have been at about a trillion barrels for the past 15 years or so. The amount of new reserves discovered each year roughly equals the amount used. I took an energy class as an undergraduate back in 1980, and I remember my professor telling me at that point that, given the world oil supply and the world oil use, our oil, at that rate, would run out in 7 years. Okay, that would have meant 1987. Obviously, our oil didn't run out, and now it's later, and we have even more years projected to use up our current reserves. So clearly we have been able to find oil at a rapid rate. That is an important part of the whole equation. However, it is impossible for us to tell how much oil is actually in the ground. It's just the same situation with minerals and metals. There's a very big difference between reserves and resources.

Now, the United States uses, currently, about one-fourth of all the world's oil, about 8 billion barrels a year, but it only has about 2%, about 21 billion barrels, of the world's reserves. The U.S. oil supply would, therefore, last less than 3 years without foreign oil, and that includes Alaska and everywhere else. So, you can see we're in a very politically unstable situation, and not surprisingly, it drives much of the United States' foreign policy.

In general, distribution of petroleum throughout the world is very uneven, much more so than many other natural resources. Almost all of it is in the Middle East. If you look at just five countries—Saudi Arabia, Iraq, the United Arab Emirates, Kuwait, and Iran—those five countries have two-thirds of all the world's oil. Saudi Arabia alone has 25%, one-quarter. Saudi Arabia is a country with a land area that is one-third of 1% of the earth's surface and a population that is 0.4%, and yet, it has 25% of the oil. So, it's not surprising that the last two U.S. wars have been fought in this region. Remember that I talked about how U.S. military camouflage fatigues have switched from jungle green to desert brown.

After those five countries, the sixth country in order of the amount of oil reserves is Venezuela. It has the largest reserves of any non-Middle Eastern country, and it was, incidentally, also the founder of OPEC.

Previously, when I talked about seismic waves, I talked about how oil and natural gas are discovered with the use of seismic imaging, and how important that was for the whole world economy. We can see where the rock layers are saturated with liquid and gas petroleum, and this greatly reduces the cost of oil in general by preventing the construction of too many dry wells. You can just drill where you know there is going to be oil, for the most part. However, it also means that we are not missing any big oil fields. The earth's crust has been very well examined from seismic imaging in all parts of the world and will continue to be so. But we are not going to get any big surprises with huge oil reserves, which means the oil is going to run out sometime.

Most petroleum is in liquid form, but there are other sources for it. Tar sands and shales have also begun to be mined, and they contain raw kerogen. Remember, that's the waxy substance that I talked about in the last lecture that has to break down in order to make petroleum products. This raw kerogen has to be mined, and it is much more expensive. However, it will become economically profitable when the price of oil goes up enough. Places like Alberta in Canada already strip mine these tar sands and cook out the kerogen for use in petroleum production.

In general, though, because oil is liquid, the ease of taking that liquid out of the ground and then converting it into gasoline makes it the fuel of choice when you need to get around someplace, when mobility is important. As a result, most of the world's transportation uses liquid petroleum. And that is another part of the whole equation. That is, different fuels have very different applications and needs in very different instances.

Another major source of hydrocarbons, as I mentioned, is that of natural gas. By the way, a hydrocarbon simply means a compound of hydrogen and carbon. Natural gas is primarily methane, which has a formula of CH_4—1 atom of carbon to 4 atoms of hydrogen—but there are also other hydrocarbon gases like ethane, and butane, and propane that are contained within natural gas. Also, natural gas, when it is pulled out of the ground, contains other non-hydrocarbon gases, and these gases have to be removed before the methane can be used for human consumption, for fuel.

Incidentally, one of those gases that is removed is helium. It turns out that about 7% of natural gas that is pulled out of the ground contains helium. Did you ever wonder where the helium comes from? Remember that when I talked about the formation of the earth, I said that Earth's gravity was too weak to hold onto it. There's no helium in our atmosphere; it just flies off into space. Remember that helium is a noble gas. It's one of those happy atoms that doesn't bond with anything else, unlike hydrogen. So how would you ever make a helium balloon?

Well, remember that I also talked about how it's the radioactive decay of the uranium and thorium that heats our mantle and drives mantle convection and drives plate tectonics. Well, one of the byproducts of that decay is helium, so helium is continuously produced within the rock of the earth by this radioactive decay. Most helium rises to the surface and flies right off into space, but some gets trapped beneath the same capstones underground that trap oil and natural gas, and it accumulates there.

Now, because natural gas is a gas, it is easily piped, and compressed, and even liquefied. It is a very important energy source for heating buildings. Natural gas actually used to just be burned off. If you ever see a refinery or a drilling process where there is a flame at the top, they are burning off the natural gas. When people were drilling for oil, they just viewed it as an unnecessary byproduct. Now it's usually collected as well when you are drilling for oil, though sometimes it is actually pumped back into the ground to create added pressure to pump out more liquid oil.

The world reserves for natural gas are large, about 180 trillion cubic meters. But we use natural gas at a rate of about 2.6 trillion cubic meters per year. Again, do the math. That works out to be about 70 years of our known natural gas reserve. But, as with oil, more is found all the time, so our resources will last longer than that. But we don't know how much longer.

Most of the world's natural gas reserves are actually in Russia, Iran, and Qatar. The U.S. has about 5.7 trillion cubic meters of natural gas, and we use it, largely for heating, at a rate of about 600 billion cubic meters per year. So, the U.S.'s known supply would only last about 9 years without imports. And there are cases where the U.S. regularly runs into shortages of natural gas, particularly in the winter, when it is used rapidly for heating. So, when your heating bill sometimes rises dramatically in the winter, that is due to a shortage of natural gas.

The other major source of hydrocarbon that I talked about in the last lecture is coal. It's fossilized swamp. And it, unlike oil, is both more abundant, and much more evenly distributed. There are hundreds of countries that mine coal within their borders, and the known world coal reserves are on the order of about a trillion tons. That's actually easy to remember, right? Oil is about a trillion barrels; coal is about a trillion tons. The world consumes coal at a rate of about 5.5 billion tons per year, so the known reserves would last about 180 years. However, the estimated coal resources are probably much larger than this and will probably extend the lifetime of coal to many centuries.

With coal, there is some good news for the U.S. The U.S. has some of the largest coal reserves of any nation—275 billion tons—more than one-fourth of the global reserves. The U.S. uses coal at about a billion tons a year, but its known supply would last about 275 years at the current rate.

And coal is well distributed. Russia has 16%, China has 11%, India has 8%, Australia has 8%, Germany has 7%, and South Africa has 5%. But interestingly, I mentioned China. China is actually the largest consumer of coal; it uses 1.3 billion tons a year, but that rate is also rapidly increasing. Coal is usually used for large industrial and electrical uses—electrical power plants. It used to be used a lot for individual heating, but not so much any more.

The problem is that because coal is solid, it can only be removed by mining. This mining used to happen primarily underground, but surface strip mining is now replacing it, largely. For instance, in the United States, of the billion tons that were mined last year, 400 million tons came from underground mining and 600 million tons came from strip mining. Strip mining is cheaper, but it also has serious consequences for the surface; you grind up the surface. However, U.S. strip mining is largely done in a process that involves reclaiming the land immediately. In other words, the land that gets torn up for the coal, gets filled in, leveled, and replanted with vegetation. This greatly reduces erosion and the effects of pollution. In fact, you can even remove and store the topsoil and put that back as well.

The major concern with coal, however, is that it adds a lot of carbon dioxide to the atmosphere. Remember that I talked previously about how we put about 7 tons of carbon into the atmosphere every year? That's largely from coal, and it seriously contributes to global heating. Burning coal also adds a variety of pollutants like sulfur oxides, and nitrous oxides, into the atmosphere.

But again, there is a solution to this: scrubbing and sequestration. Chemicals, catalysts that are added to the combustion process, can absorb carbon and those other pollutants like sulfur. That material then can be taken, that carbon, and buried back under ground, sequestered away. You need to have that carbon underground for at least 1,000 years or so before it gets reabsorbed by the ground layers.

Current technologies are able to remove most of that carbon—more than 95%. Of course, there is an additional cost. The cost of removing the carbon and sequestering it away is hard to estimate, but it's probably on the order of about one-third of the cost of the production of coal itself, so that is a lot of money. But in reality, it is probably far less than the true cost of carbon emissions without scrubbing and sequestration, because you would have to account for national health costs from the pollution. Or from the global warming, you would have to account for water uses during droughts and building larger levees for rising sea levels. You look at the long term picture of taking that carbon out and putting it away, it's going to be much cheaper than dealing with the consequences of not doing that. And technologies are improving. That process will probably get considerably cheaper over time.

This is a theme that I am going to come back to in Lecture Forty-Five, but I want to say it here. I personally believe that we shouldn't be pessimistic about the problems that come up with using energy sources and a variety of other things. I mean, humans are remarkably resourceful. Look at London. It used to be black and filled with smog from the burning of coal. That doesn't happen any more. Yes, we have become the greatest agent of geologic change, and we are changing the surface rapidly, but that also means that we have the ability to handle problems as they come up, as long as we understand what the problems are and we make that understanding widely known. That is part of my job as a scientist and as an educator.

There is one source of hydrocarbons that I haven't talked about yet, and it's very exciting, very interesting. I just touched on it earlier. It's called "methane gas hydrates," also known as "methane clathrates." They are frozen deposits of methane-containing ice. If you take the stuff out of the ground, it looks like snow, but if you hold it in your hand and light it, it will burn. As the snow evaporates, there is methane actually trapped within the ice, and it will be released. It's a large source of hydrocarbons. It forms naturally in marine sediments, usually at a depth of about 1/3 to 2 kilometers deep. It also forms in permafrost and tundra regions, and it's formed usually through biogenic processes, as a byproduct of bacterial processes. It can also form from a breakdown at high temperature of deeper marine organic sediments that get trapped in shallower sediments as that methane rises up. But there is an enormous amount of it.

Now, there is a huge engineering challenge in getting it out. We haven't quite figured out how to get this snow out of the marine sediments without having it evaporate and lose the methane out into the atmosphere. But that is an engineering problem that I'm sure we will work out.

The advantage of clathrates is that it has been discovered in many regions, all around the world, and there is just a huge amount of the stuff. The initial estimates were very, very high. The amount of carbon stored in these methane clathrates was twice as much as all other fossil fuels combined. More recent assessments have brought that down, somewhat, but it's still likely that these clathrates contain about half as much carbon as all other fossil fuels. That's a huge amount. Again, it's hard to tell how much is actually there. We've just begun to search for it. Nonetheless, it's quite possible that clathrates contain ten times as much carbon as natural gas resources, and one-third to one-half of the world's total fossil fuel carbon budget.

However, there is also a growing sense that methane gas hydrates play a very important role in rapid climate change. Remember many of those sudden increases in global temperatures I talked about? Like what happened at the end of the Snowball Earth episodes back in the Precambrian, when the earth swung rapidly from being totally frozen to a runaway greenhouse effect that overheated the earth. At the end of each of these small, recent Ice Age episodes, where the temperature got colder and colder and then rapidly heats at the start of one of these interglacial periods—these are probably due to the rapid release of these frozen methane clathrates. When it's cold, the sea level drops. These ocean sediments are exposed, they erode, the methane gets emitted, you trigger off a runaway greenhouse effect. So, we obviously need to keep a very close track on these clathrates. They have the potential to release a tremendous amount of methane in a hurry, so we need to be careful we don't trigger one of these large releases.

The last area of nonrenewable energy is in the form of nuclear fission. Nuclear power is not a fossil fuel, but it is a nonrenewable energy source because we have limited amounts of the elements of uranium and plutonium. Nuclear fission involves the controlled release of energy from the splitting apart of either uranium-235 or plutonium-239 isotopes. This is totally different from the nuclear fusion that I talked about, like what goes on within the sun, where you smash hydrogen atoms together to make helium, releasing energy in the process. Here, you break apart large atoms, but also, in the process, a small amount of mass is converted into energy.

Currently, about 15% of the world's electricity is generated from nuclear fission. About 21% of the United States' energy comes from nuclear power, generated currently by 103 nuclear reactors at 64 different power plants. But if you go to some places in Europe—for instance, countries like France, Belgium, and Lithuania—they get more than 50% of their electricity from nuclear power.

During the nuclear fission process, a uranium isotope gets bombarded by a neutron at very high energy, and if the neutron has enough energy and hits that uranium atom, it breaks apart, forming two other smaller atoms (krypton and barium) and two highly charged neutrons, which then go flying off. In the process, a small amount of mass gets destroyed. Remember $E=mc^2$? You take a small amount of mass, you convert it into electromagnetic radiation, and you get a whole lot of radiation, a tremendous amount of energy.

The process however of having those neutrons fly out gives you a chain reaction. Usually in a nuclear reaction, the uranium gets stored in rods, and those rods are submerged in water. The heat gets absorbed, and the chain reaction stops, and you can therefore control the chain reaction. It's called a controlled reaction. The heat that gets radiated from this goes into heating water and running turbines to make electricity, just like any other turbines, only it is incredibly efficient, given the amount of fuel involved. One kilogram of uranium, 2.2 pounds, produces as much energy as 1,500 tons of coal.

Most nuclear fission uses uranium-235. Unfortunately, that is not very abundant. Uranium-238 is more than 140 times as abundant. However, using a special type of reactor called a "breeder reactor," you can convert uranium-238 into plutonium-239, and that can be used for nuclear power as well.

The world reserves of uranium oxide currently are at the order of about 3.5 million tons. Our resources are probably 10 million tons. The world consumption rate is about 75,000 tons of uranium ore per year. And again, you do the math. That gives us 100 to 150 years of uranium at the same rate that we are using today.

However, interestingly, it might be possible to extract uranium out of seawater and also other minerals as well, and if the price of energy rises high enough, at some point, that might become economically feasible. That might happen sooner than later. The price of uranium has risen quite rapidly. It was at about $10 a pound in 2003. It has now reached over $110 a pound in 2007.

There is a large upside of nuclear fission, and that is that it doesn't produce any greenhouse gases. There is no carbon released. There are serious objections that people have made to nuclear power on both political and environmental bases. For instance, a large concern with nuclear fission has always been a reactor meltdown, like the China Syndrome, the sort of thing that happened at Three Mile Island in Pennsylvania in 1979, or the Chernobyl reactor in the Ukraine in 1986. Part of the reason was that when people moved early on to nuclear power, they really rushed early designs. Reactor designs are now much more efficient and much safer. However, currently there is no demand for them in the United States, while coal remains so cheap. Also, there are risks of a reactor meltdown in the context of global terrorism. If nothing else, it makes for a lot of movies and TV shows.

There's a more important concern, a larger concern, and that is that nuclear fission produces highly radioactive byproducts, and they have to be disposed of and left alone for extremely long periods of time. Over the past 20 years, it is estimated that the U.S. has created more than 8 million tons of radioactive waste, and there is still no centralized location to put them. Most radioactive wastes are stored onsite at the reactor, currently indefinitely. That's usually fine, but in some cases, containers leak, and this material does get out into the environment. For example, a few miles north of the Arch in St. Louis on the Mississippi River is a building that is permanently sealed away. It was used during the Manhattan Project to create fissionable material for the nuclear bomb. People using it developed very high cancer rates. It is still tremendously hot, and no one goes in or will go in for—I don't know—indefinitely. There is nothing you can do with it.

These sorts of wastes have to be dealt with at some point. There is a lot of debate and discussion as to what to do with them, where to put it. Do you put it on the bottom of the ocean seafloor? Do you blast it into space? We have partially constructed a disposal facility at Yucca Mountain in Nevada, but currently, at this point, it is still locked in a legal debate as to whether or not we'll actually be able to put our wastes there.

There are also other issues. You would have to have many trains carrying radioactive waste all across the country to a repository like Yucca Mountain. Suppose there was an accident with that train? Another issue is that plutonium is incredibly toxic. Even the tiniest microscopic amount of plutonium will kill you, and these sorts of issues have to be figured out and taken into consideration.

So, in this lecture, I've talked about the two major forms of nonrenewable energy sources: petroleum and uranium. They power the major part of our world's economy, and they are going to be around for our lifetimes, and our children's lifetimes, and our grandchildren's lifetimes. But over time, the increased problems of dealing with these byproducts, of carbon and radioactive waste, and the dwindling resources, will make them less attractive. But that's okay, because in the next lecture, I'm going to discuss a whole suite of energy sources that are ready to take their place.

Lecture Forty-Four
Renewable Energy Sources

Scope:

Humans will soon get almost all of their energy from solar-driven sources. The reason is simple: There is so much of it available, and it won't run out for many billions of years. Earth receives 174,000 TW (terawatts or trillion watts) from the sun. This is about 10,000 times the human energy use of 15 TW. Solar energy comes in many different forms: active solar (using silicon-based solar panels), passive solar (using the heat from sunlight), wind power (using sun-driven atmospheric circulation), hydroelectric power (using the sun-driven water cycle), and biomass fuels (using sun-driven photosynthesis). All of these energy sources are becoming more attractive as technologies improve and the prices of fossil fuels increase. There are also some renewable energy sources, such as geothermal and tidal, that do not rely upon sunlight. Portability of some solar energy sources is problematic, in particular, for use in cars, but battery and hydrogen fuel cell technologies are being developed to address this.

Outline

I. This lecture will address renewable energy sources.

    A. The amount of energy we get from the sun is relatively unlimited and essentially never-ending.

    B. The Earth receives more solar energy every hour than the total amount of energy that humans use in a year.

    C. Earth receives about 174,000 TW from the sun—10,000 times the total human energy use of 15 TW.

    D. We have to figure out how to convert the sun's energy into forms we can use in our daily life.

        1. We need electricity to power lights.

        2. We need heat to stay warm.

        3. We need portable energy for transportation.

II. Electricity is a wonderful form of energy because it is easily movable due to the equivalence of electricity and magnetism.

A. One of the four fundamental forces of the universe is the electromagnetic force. A turbine creates electricity from magnetism by spinning large coils of wire through a magnetic field.

B. Water is usually used to spin the coils—either through flowing water or expanding steam.

C. The big concern is efficiency. Energy conversions always include some loss of energy to heat.

III. Solar energy is usable in both direct and indirect forms.

A. Direct solar energy is the use of solar radiation to create electricity using photovoltaic cells and steam turbines or to directly heat water or buildings.

B. Indirect solar energy involves the natural conversion of solar energy through geologic processes.

C. There are many benefits of using solar energy, in addition to the fact that there is so much available.

1. Solar power is pollution free; there is no release of carbon dioxide, no acid rain, and therefore no clean up.

2. Although initial expenses can be high to set up a solar power facility, operating costs are generally low because they tend to run without much intervention or maintenance.

D. Solar power comes in several forms and can be adapted to different climates and different regions.

E. Energy futures are highly dependent upon government and corporate policy.

IV. Direct solar power uses sunlight for heating or electricity.

A. Passive solar power is using sunlight directly for heat and involves designing homes and buildings to take advantage of the sun that hits the building. Active solar power is the conversion of sunlight into electricity.

B. One challenge here is the efficiency of solar panels.

C. In North America, the land receives an average insulation of about 125 to 375 watts per square meter. If you took a square region of 600 kilometers by 600 kilometers, you could generate all of the world's energy needs.

D. Energy demands are increasing, but silicon solar panel technology is improving, and we have no shortage of silicon on the planet!

E. To maximize efficiency, it is best to create power locally and distribute it over small distances; active solar power also works well in areas that are off the grid.

F. Large power plants that generate electricity to power large regions use large reflecting mirrors to concentrate sunlight, heat fluids, and drive turbines.

G. We could eventually get more energy with solar collectors put into orbit in space.

H. One day we may need more than 10,000 times the energy we use now, and we have that power in the form of the sun.

V. We harness a small portion of the work water does as it washes off the land as hydroelectric power, which is currently the most utilized solar energy source.

A. Advantages of hydroelectric power are that it is inexpensive and clean.

B. Disadvantages of hydroelectric power include a loss of water to evaporation, an increase in mosquito-borne diseases, potential interference with animal and fish migration, and area limitation to regions with large streams and steep slopes.

C. Hydroelectric power works best in temperate and tropical regions. The largest producers of it are China, Canada, Brazil, and the United States.

VI. Wind power is another form of solar power. This is the energy source with the most rapid advances in technology.

A. Only about 1% to 3% of sunlight actually goes into wind; however, that is still a lot of energy.

B. The strongest winds occur high up in the atmosphere; however, there are places on Earth's surface such as the plains states, mountainous regions, and shorelines where the wind regularly blows quite strongly.

C. Winds are often strong at near-shore or offshore locations because of the large changes in temperature there.

D. The advantage of the wind turbine over the old windmill is that technological engineering advancements, such as allowing it to rotate more slowly, have made it more efficient, more durable, and less of a threat to birds.

E. The largest wind turbines, in ideal conditions, currently each generate enough electricity to power about 160 homes. Recent studies indicate that wind power could be scaled up with today's technology to provide five times the current global energy use and 40 times the current electricity requirements.

F. Wind power is clean and inexpensive to maintain.

G. New technologies are being developed to harness the energy in another form of energy that comes from wind: ocean waves.

VII. Biomass is another way to convert sunlight into energy (e.g., through photosynthesis).

A. Biomass is the use of plants to produce biochemical energy. We currently can convert certain crops into biofuels.

B. In the long term it may not be sustainable. We may need crops to feed an increasing human population.

C. Garbage is another source of energy. Biogas is methane extracted from waste treatment plants.

VIII. There are two major sources of renewable energy that are not powered by sunlight: Earth's heat and the gravitational pulls of the sun and moon in the form of tides.

A. Geothermal power can be used in volcanic regions where shallow drilling will allow access to the hot rock that will provide steam for turbines.

    1. The largest single steam field in the world comes from The Geysers in California.

    2. Iceland gets more than 50% of its electricity from geothermal energy.

    3. On the island of Heimay, people heat their houses with geothermal energy from the same volcanic processes that almost destroyed them.

B. Tidal power can be tapped to run turbines and generate electricity. Tidal swells can be more than 10 meters in some places, and you can use them to run turbines. It is very efficient, but limited to just a few locations.

C. One form of renewable energy that was once considered the likely source of future energy is nuclear fusion; however, we have not been able to re-create the process of fusion in an economically feasible way.

D. Portability of energy is a big concern. One option is using hydrogen fuel cells, which are not a source of energy but a means of converting it from one stationary source into a portable form that can be used in cars and trucks.

E. Electric cars are a proven technology—they work well and there is a huge demand around the world for them.

IX. If we play our cards right, there is no reason humans cannot be around for hundreds of millions of years.

A. The foundation of this will be to find a sustainable way to live. An important component of this will be renewable energy from the sun that is clean, efficient, and limitless.

B. The only questions will be how long it takes to move to a sustainable lifestyle and how smooth will that transition be.

Recommended Reading:

Flavin and Lenssen, *Power Surge: Guide to the Coming Energy Revolution.*

Hinrichs and Kleinbach, *Energy, its Use and the Environment.*

Questions to Consider:

1. Given that we have a growing population that is on the verge of starvation in many parts of the world, how sustainable do you think it is to grow crops to burn in cars and planes?

2. Given the part of the world you live in, what do you think would be a logical mix of renewable energy sources for your region?

Lecture Forty-Four—Transcript
Renewable Energy Sources

Welcome. In this lecture, I want to talk about renewable energy sources. Now, in the last lecture, I talked about how fossil fuels are going to dominate the global energy budgets for centuries to come until they are nearly all used up. But there is every reason to be optimistic about the future of society and of our economy.

Our society is not going to collapse for a lack of energy for a very simple reason. Go ahead, walk outside during a sunny day. What do you feel on your face? That's heat, energy from the sun. The amount of energy we get from the sun is relatively unlimited, and essentially never-ending, at least from a human perspective. The earth receives more solar energy every hour than the total amount of energy that humans use in a year.

Earth receives about 174,000 terawatts (trillion watts) from the sun—or, about 120,000 terawatts of useful energy once it reaches in through the atmosphere toward the surface. That's 10,000 times the total human energy use of 15 terawatts.

Let's stop for a moment and think about what sunlight is. It is electromagnetic radiation, waves. Remember, waves are the transmission of energy, so the sunlight from our star is energy that is being transmitted from the nuclear fusion occurring in the core of the sun to the surface of our planet. Remember, energy was that thing that was hard to define but that had the capacity to do work. And energy also obeys the law of conservation that when it transformed from one form to another, the total amount had to be conserved. Remember, mass was part of that; mass was a form of energy.

So, here's the trick. This energy in the form of sunlight comes in to Earth's surface, and we have to figure out ways to get it into useful forms that we can use in our daily life. Well, what is it that we need? Well, we need electricity to power lights, and motors, and all sorts of things. Motors take the electricity, which is in the form of flowing electrons, and will make a wheel spin. Well, if it will make a wheel spin on an axle, you can do almost anything else with it with gears and pulleys. Once you have things spinning, you can supply our entire industrial needs—factories, mills, treadmills, whatever.

What else do we use energy for? Well, we need heat to stay warm. Usually, we use natural gas for that, but heating can also be done with electricity. We need portable energy for transportation, what gasoline does for us with cars and trucks, though, interestingly, electricity can also be used to do this. In fact, in many forms, it already is. Subway trains and trolleys already use electricity for transportation. And of course, we have the option of portable batteries as well.

Electricity is a wonderful form of energy, because it is easily moveable because of the equivalence of electricity and magnetism. Let's go back to the basic, fundamental forces of the universe. I talked about the four forces, and one of them was the electromagnetic force. Well, the reason that electricity and magnetism are listed as one force, the electromagnetic force, is because the two can transfer back and forth. If you take a wire of metal and move it through a magnetic field, you generate an electrical current within that wire. You create electricity from magnetism, and that's what a turbine does. It takes large coils of wire, spins them through a magnetic field, and generates electricity.

Well, how do you spin the coils? What we usually use is water, either water flowing downhill, running a hydroelectric power plant, or water turning into steam that expands and is used to spin those coils. So, you need to heat water to make steam to create a turbine to create electricity. That electricity then flows through a set of wires, into a motor, which creates another spinning wheel again. The electricity is a way to take a spinning wheel in one place, transfer that energy, and make a spinning wheel in another place. It's very flexible.

There is a big concern in all this, however, and that is efficiency. Energy conversions always include some loss of energy into heat, from friction, radiation, conduction, etc. You can never convert all your energy from one source into another useful source, something that we always have to consider whenever we are dealing with energy in any of its forms.

So, we have all this electromagnetic energy from the sun. The trick is how to convert it cheaply and efficiently into other forms we can use. Solar energy is usable in both direct and indirect forms. The direct solar energy is the use of solar radiation to create electricity, as I've talked about before, either using photovoltaic cells or other means, which isn't usually very efficient, or directly, as in directly heating water or buildings. And that tends to be very efficient. The other form of solar energy, what we call indirect solar energy, involves the natural conversion of solar energy into all sorts of other different forms through geologic processes. I will get to these in a bit. In addition, there are also some renewable energy sources that don't come from sunlight, like geothermal energy or tidal energy, and I'm going to talk about these as well.

Now, there are many benefits of using solar energy, in addition to the fact that there is just so much of it available. First of all, solar power is pollution-free. It doesn't contribute to global heating, because there is no release of carbon dioxide. In fact, there are no other pollutants as well. There is no acid rain generated from sulfur or nitrous oxides. So, solar power, as a result, is also inexpensive that way. There is no cleanup required afterwards. Also, facilities using solar power tend to run with very little intervention or maintenance once they're set up. The initial expenses can be quite high, and that's a downside of solar energy compared to simpler fossil fuel powered turbines. But the operating costs are generally quite low compared to nonrenewable fossil fuels over the long operating lifetimes, because there's not a lot of maintenance. The whole process runs quite cleanly. So, in the long run, using solar power is already economically advantageous in many situations.

Another advantage of solar power is that indirect solar power comes in several different forms. It can be adapted to the different climates of different regions around the surface of the earth. Again, because renewable energy sources are already cheaper than fossil fuels in some cases, technologies are in the process of developing to take full advantage of them.

The actual path that energy futures will take turns out to be highly dependent upon government and corporate policy. Government incentives like California's programs of the Solar Initiative and the Million Solar Roofs program are going to go a long way towards helping renewable energy sources develop more quickly, easing us in that transition from fossil fuels to something else.

Let's start with direct solar power, which uses sunlight for heating or electricity. As I said, it comes in two forms, active and passive.

Passive solar power is using sunlight directly for heat. It is a very good local means of reducing energy needs, because homes and buildings can be designed to use sunlight directly to supply much more of their heat. Simple things, common sense things—where you put your windows, the angle of your roofs, or even setting up systems of pipes that let water flow through the roof to heat up and take advantage of the sun that hits the building. In fact, even in third world countries, simple solar ovens, which are really just a set of folding out reflecting sheets, are replacing scarce firewood as a means of cooking food. That takes the load off strained vegetation in many parts of the world that are especially suffering from desertification. And it also reduces carbon dioxide production going into the atmosphere.

Active solar is the direct conversion of sunlight into electricity, and it's used on both local and large scale power plant levels. The real challenge here is the efficiency of solar panels. That efficiency is really key in active solar, and it's been considerably increasing over time. For instance, currently, commercial panels are about 15% efficient. In other words, 15% of the sunlight that hits the solar panel actually gets converted directly into electricity. Experimental panels now bring that number up to about 22%, and there are now technology panels that are being developed that have efficiencies of over 40%. That is really important, because if you have double the efficiency, you need half as many solar panels.

In North America, the land here receives an average insolation (energy from the sun) of about 125 to 375 watts per square meter. If you take the current 15% efficiency of photovoltaic cell conversion, that gives you 20 to 55 watts per meter squared. That's about one 40 watt light bulb being generated for each square meter of land. That doesn't sound like a lot, but there is so much land on Earth. If you took a square region just 600 kilometers on a side (400 miles by 400 miles), you'd generate all of the world's energy needs. This is something that is smaller than a thousandth of the world's surface and would give the whole planet its energy needs—all the humans on the planet. That's really remarkable.

Energy demands are increasing over time, but so is silicon solar panel technology. And silicon is not something that we are going to run out of on this planet. There is a lot of silicon. It is the fundamental material out of which silicate rocks are made, so that's a resource that we don't have to worry about.

Of course, you never would actually want to have all your energy produced in one place. There would be a tremendous loss of efficiency, because as that electricity travels along power lines, the power lines radiate off heat, and you would lose energy. So, you actually want your power created locally and distributed over small distances. As a result, active solar power actually works very well in areas that are off the grid, that are far from large power distribution, because solar panels can come in many sizes. You can generate enough power for a single house as well as a whole block or town.

There are large power plants that do generate electricity to power large regions. They don't usually use solar panels. They actually use sets of reflecting mirrors and a variety of different structures to concentrate sunlight, and heat fluids, and then drive the turbines to generate electricity. Some of the most successful and cost effective solar power plants actually use a structure that involves a parabolic trough. It's a long, curved sheet that generates a reflected light onto a pipe in the middle. That pipe contains a fluid that heats up and drives the turbines. One of these in California, the largest such plant, actually generates 350 megawatts (350 million watts) of power from one active solar power plant.

Now, if we ever needed to, we could even get more energy by putting solar collectors out into orbit in space the way we do with satellites. There are discussions now going on about how to collect sunlight with large space-based reflecting mirrors that focus the sunlight onto a surface based receiving area. It's still off in the future. It sounds like science fiction, but there are no physical obstacles to this, just the engineering challenge, and we've done a lot more remarkable things, like land people on the moon and actually bring them back.

Even far off in the future, there may be a day when we need more than 10,000 times the energy we use now. Maybe to mine deep down into the earth, to do things we can't even imagine now. Well, we have that power in the form of the sun, and there are actually theoretical discussions going on now about how to wrap a sphere around the solar system itself in order to capture much of the sun's light before it goes off into space! It is science fiction now, but look, the books of Jules Verne, which are not that old, are filled with things that were science fiction back then but are reality now. It may not be that far into the future.

Where else in the course have I talked about power from the sun and Earth's surface? The water cycle, the sun-driven hydrologic cycle. I've talked about how much work water does as it washes off the land. Well, we can harness a small portion of this as hydroelectric power. In fact, we already do. It's actually the most utilized solar energy source. Water flows down; it physically moves the turbines that generate electricity and provides 5% of the world's energy use. Now, 19% of the world's energy comes from harnessing that gravitational potential energy of water as it flows downstream.

Hydroelectric energy has some real advantages: It's inexpensive, and it's, again, very clean. There are some disadvantages that largely come from building dams, things I've talked about already. There is a loss of water to increased evaporation when you let it sit out in a wide area. There are increases in mosquito-borne diseases. Dams also can prevent migration of animals like fish up and down stream along the rivers, though many dams now have special spillways that allow the fish to bypass the dam. Hydroelectric power is also going to be limited to regions with large streams and steep slopes, because you actually need a big drop in elevation in order to run large turbines. But hydroelectric power has been used for a long time. The first plant actually was built in 1870 in Cragside, England. Hydroelectric power in general, however, works best in temperate and tropical regions, so the largest producers of hydroelectric power are China, Canada, Brazil, and the United States. In the U.S., 80 gigawatts (80 billion watts) of electricity is generated from rivers.

Well, where else on the surface does the sun's energy go? It goes into the atmosphere, into winds. Wind power is another form of solar power. Currently, wind power generates less than 1% of the world's electricity, but it's also the area with the most rapid advances in technology. Wind power actually quadrupled simply over the period between the years 2000 and 2006. A small percentage of sunlight actually goes into wind. About 1–3%, depending upon the conditions for solar radiation, entering the atmosphere gets converted into wind. That's still a lot of energy.

Now, most of the winds, the strongest winds, occur very high up in the atmosphere—things like the jet stream. Up there, there are continuous wind speeds of well more than 100 miles per hour, but it's hard to get up there. Fortunately, there are also places at Earth's surface where the wind regularly blows quite strongly. The plains states of the United States are a very good location, so are mountainous regions and shorelines as well.

It's interesting: Winds are often strongest at near shore or offshore locations because of the large change in temperature that happens over the course of a day. Remember that I talked about how water changes its temperature much more slowly than land or rock simply because of its high heat capacity? Well, during the daytime, the land may be warmer, the air might be rising, and you get wind coming from the sea onto the land. At nighttime, the land temperature cools, and you have wind blowing from the land out to the sea. If you have wind turbines, you can change their direction, and they could take advantage all day and night of the wind's blowing along shore. A lot of wind turbines are increasingly being located along shorelines.

Yes, the downside of windmills, of wind turbines, is that you are going to have to see a lot of windmills, a lot of wind turbines, along the surface of the land, but remember, the appearance of windmills is nothing new. They have been used for centuries. They are an icon, for instance, of the Dutch landscape in Europe. Remember Don Quixote tilting at windmills? I think that we will just get used to seeing wind turbines as well.

The advantage of the wind turbine over the old windmill is that there are really important technological engineering advancements that have allowed them to become much more efficient, largely in the form of having much stronger metal alloys that don't fall apart, and systems that involve the blades spinning more slowly but still being able to generate large amounts of electricity. Lower spin velocities and stronger alloys means that they are more durable; they last longer. And also, they pose less threat to birds, especially migratory birds, which are damaging to the wind turbines and to the birds.

Currently, the largest wind turbines, in ideal conditions, can actually generate about a megawatt of electricity each. That's enough to power 160 homes. You can also have small rooftop mounted turbines on the order of about a kilowatt, and that's increasingly used, because you can go to any remote location, particularly in a mountainous region and, again, generate electricity right on the spot. Many farmers in the U.S. Midwest region are currently putting in turbines on their land in order to increase and supplement their farm income. Often they actually make more money per acre from selling electricity to the power grid than they do from selling their crops. In fact, recent studies have shown that with current technologies, wind power could easily be scaled up to provide more than 70 terawatts of power. That's five times the current global energy use and 40 times the current electricity requirement.

And again, wind power is clean. There is no pollution generated, and once they are constructed, they're very inexpensive to maintain. There's no fuel that you have to buy. The source comes from the wind, which just is continuously there, day after day, year after year. Wind power also can be scaled up in the future. Giant, airborne wind-generating kites have been planned that would take advantage of those strong steady winds way up high. They haven't been constructed yet, but they're actually closer to reality than putting giant solar mirrors up into space.

There's one other form of energy that comes from wind, and that's the generation of waves on the ocean. As the wind blows across the ocean's surface, it creates waves. Those waves reach the land, and new technologies are being developed to actually harness the energy in those ocean waves. There's a device that was constructed off the coast of Portugal in 2006 called a "Pelamis machine," which actually takes energy from those waves and converts it into electricity. It's small, about a little but more than 2 megawatts of electricity, but again, this is a new technology that may build up over time.

Let's think of where else sunlight goes. It goes into the biosphere. Biomass—that's another way to convert sunlight into energy, and that wasn't something that we had to figure out. Photosynthesis—life figured that out on its own, 2.5 billion years ago. Biomass is the use of plant crops to produce energy. Now, of course, biomass is the oldest source of energy (burning wood), and it's actually still used in many parts of the world. But we now can actually grow certain crops and convert that biomass into biofuels, into liquid hydrocarbons, like ethanol. So, crops like corn, sugar beets, sugar cane, and switch-grasses can be converted into biofuels, and that is already taking some of the burden off of fossil fuel demands. In the long term, this may not be sustainable. It may not be an ideal source of energy, because with the increasing human population, we may need that output from limited agricultural croplands to feed people rather than feeding cars.

Talking about increases in the number of people, there's another wonderful source of energy, and that comes from garbage. Biogas is methane that's extracted from waste treatment plants of crops, sewage, paper production, and animal waste. I mean, bacteria are going to eat that stuff anyway, so we may as well harness it and capture that methane and put it to use. By the way, it's interesting: Kerosene is currently the only major source of fuel that is usable for jet airplanes, and that's a real problem as our fossil fuels run low. It turns out that an acceptable replacement can actually be produced from biomass.

Now, there are two major sources of renewable energy that are not powered by sunlight: Earth's heat and the gravitational pulls of the sun and moon in the form of tides. Now, both of these have very limited geographic applications, and aren't going to be a large part of our energy budget. But they are used very importantly in some parts of the world.

Earth is cooling down—I've talked about this quite a bit during the course—at a rate that is currently three times the human energy consumption, but it does so very slowly and broadly. Only about 80 milliwatts (80 thousandths of a watt) of power come out of the surface per square meter. Normally, you have to go down several kilometers under the surface of the Earth in order to find water that is more than 100°C in order to turn it into steam to drive the turbines to get electricity. But geothermal power can be used in those areas where you only have to drill into shallow regions of the rock to reach very hot rock that will give you the steam for your turbines.

That usually means volcanic regions. Currently, about 100 gigawatts of thermal power is generated by tapping into these hot regions underground. The largest single steam field in the world comes from a place called The Geysers in California. Remember that we've got the Juan de Fuca Plate subducting beneath North America, causing magma to rise up to the surface, and we've got all these volcanoes from Lassen all the way north up into Oregon and Washington. Well, that volcanic rock can be tapped into in order to give us power.

Iceland is the ideal location. Iceland gets over 50% of its electricity from geothermal energy, and also a tremendous amount of direct heating of homes. Remember the island of Heimaey, where they stopped the flow of lava? Well, they not only got a better harbor out of that, but once the lava cooled, they also had a tremendous amount of heating, and they warm all their houses with that lava, that rock that's quite warm.

Tidal power uses the conversion of gravitational potential energy causing water to flow kinetically at the surface. It can be tapped into to run turbines and generate electricity. It uses the twice daily raising and lowering of sea level. It's just like hydroelectric power, only it can be reversed to handle water flowing both into a region, like a river or fjord, and then back out again, twice a day.

In some parts of the world, if you have just the right geometry of bays and channels, the tidal swell can be more than 10 meters. Remember that I talked about the Bay of Fundy, where it was 50 feet up and down in ideal conditions? Well, you can tap into that flow of water to run turbines, and like hydroelectric power, it's very efficient (about 80%). However, it's limited to just a few locations around the world.

One possible source of renewable energy that was once considered to be the likely source of most future energy, but just hasn't panned out, is nuclear fusion. This involves taking heavy hydrogen atoms and smashing them together. In that process of fusion, energy is released. That's what goes on in the core of the sun. However, we haven't been able to recreate that in an economically feasible way yet, because it takes a lot of energy to smash those atoms together, and the best that we've been able to do is get the same amount of energy back that we've put into smashing them together. Nuclear fusion may still work one day, but it's not going to play an important role in the near future.

As I mentioned, the portability of energy is a really big concern; transportation accounts for a huge percentage of the energy that we use. One option is using hydrogen fuel cells, and people sometimes get confused by this. Hydrogen fuel cells are not a source of energy. They are simply a means of storing energy. They're like batteries. They are means of converting electricity from one stationary source—a big coal plant, or a nuclear, or direct active, or solar, or wind plant—converting that energy into a portable form that can be used in cars and trucks. It very well may one day help replace petroleum as a form of mobile energy. What happens with a hydrogen fuel cell is that you take the energy in one place to break apart water into the separate oxygen and hydrogen atoms, and that hydrogen is pumped into battery-like containers within a car. Another approach is simply to use a battery and have an electric car, where you draw electricity more directly, right off a grid. In other words, you take your car, you drive it home, and you plug it in the wall, just like you would your cell phone. Electric cars are already a very proven technology. They work remarkably well, and there is already a huge demand around the world for them.

Now, over the past two lectures, I've talked about the two sides of energy, as far as human needs are concerned. One part, the last lecture, represented the world of today, largely driven by fossil fuels, and the other part that I've talked about here, represents the world of tomorrow and of many tomorrows to come. It's solar power that has practically no downsides to it.

Now, I can't help it—I'm a geologist—I take a long view of things. Yes, human beings are the new kids on the block. We've been around for merely a couple hundred thousand of years. That's insignificant in the long geologic time scale, but if we play our cards right, there's no reason it can't be hundreds of millions of years for us to be here. I mean, after all, there are a few species that have managed to do that. The foundation of this will be to find a sustainable way to live, and an important part of this will be renewable energy from the sun which is clean, efficient, and, as far as we're concerned, limitless. I mean, the sun is not going to go red giant for another 4 billion years. But it's inevitable, looking at the big picture. There is just such a small amount of fossil fuel on the planet. The only questions will be how long it takes to move to that sustainable lifestyle, and how smooth that transition will be.

In the next lecture, I am going to talk more about our ability to make this transition. I'm going to talk about our recognition and understanding of the immense power we have become as a geologic force.

Lecture Forty-Five
Humans—Dominating Geologic Change

Scope:

We are an integral part of Earth, continuously sharing our atoms with the rest of it. We have also become the most significant agent of geologic change. The amount of paved surfaces in the United States now exceeds an area greater than the state of Ohio. We have altered rivers, increased cloud cover, and moved mountains to get at the mineral reserves underneath. We have tripled erosion rates with our agriculture and deforestation, and filled the space around Earth with the debris of our space missions. The release of carbon dioxide from fossil fuels and methane from our cattle is raising global temperatures, changing climates, and causing global sea levels to rise. Human activities are causing the extinction of enough species such that this century will mark the end of the 65-million-year-old Cenozoic era.

Outline

I. There is nothing new about life altering the planet. What is remarkable is the speed with which one particular species, us, humans, are doing this.

   A. There are many previous examples of life significantly changing Earth's surface.
   1. Photosynthetic cyanobacteria 2.5 billion years ago removed carbon dioxide from the atmosphere and produced oxygen.
   2. Ocean worms ended the Snowball Earth runaway greenhouse effect by churning up marine sediments.
   3. Corals and shells created limestone.
   4. Leafy trees became a massive store of carbon dioxide, buffering the climates.
   5. Single-celled bacteria and archaea catalyze almost all surface geochemical reactions.

   B. What is unusual is that we have managed to significantly alter our planet in just 200 years through a scientific understanding of how the earth works and through the development of this information with engineering and industry.

C. It may sound like a litany of bad deeds, but humans are a young species like children, and like children we make mistakes. Each mistake, however, is an opportunity to teach and to learn.

    1. We passed legislation that stopped the production of chlorofluorocarbons, and the growth of the ozone hole has largely stopped.

    2. In 1973 the United States began to phase out lead in gasoline, and lead is now almost entirely gone from streams.

    3. Nuclear testing was moved underground, and now such testing is almost nonexistent around the globe.

II. We are putting a lot of junk up into space.

A. There are currently more than 600,000 objects—remnants of previous satellite launches—of at least one centimeter in size that orbit around the earth.

B. The U.S. Strategic Command tracks the location of the largest 10,000 of these objects. They pose serious hazards to the international space station, all other space missions, and all of the satellites we have up there.

III. Humans are significantly changing the appearance of many parts of the land.

A. Erosion rates have increased dramatically primarily through the removal of vegetation and increase in agriculture.

B. For the past 10,000 years humans have been removing forests, and this has had a tremendous effect on the biosphere and atmosphere. Currently about 35% of the world's lands are now used to produce human foods.

C. Each year about 70 gigatons of valuable topsoil is lost, although that number is decreasing as better agricultural practices have been developed.

D. Wetland regions have been drained or filled in, removing water from the land and local atmosphere, making regions more arid. This also causes larger seasonal temperature changes.

E. The amount of the United States that is now paved is greater than the area of the state of Ohio. This affects land temperatures, which in turn affects storm patterns.

F. Paving also affects water flow patterns because it increases the amount of water that is removed from land and put into streams. It prevents water from going into the ground—it all gets washed out to the sea. This is a huge problem in urban areas.

G. In the United States, humans generate about 450 million tons of garbage each year.
  1. Recycling has gone up quite dramatically in the United States since 1991.
  2. Places like Europe recycle an even higher percentage of their waste.
  3. This is a lesson that humanity is successfully learning.

IV. Humans have had a significant effect on streams and lakes.

A. The Clean Water Act went into effect between 1972 and 1977 and began cleaning up polluted lakes and streams.

B. Streams are now mostly clean.

C. Many streams, however, still have very high levels of pathogenic bacteria due to livestock and human wastes. Part of the reason for this is that our population is increasing and so are our food supplies.

D. Groundwater reserves in many parts of the world have been significantly contaminated.
  1. Contamination comes from a variety of sources: industry, dumping, petroleum spills, and agricultural chemicals in farming areas.
  2. It may be thousands or tens of thousands of years before the groundwater recovers from the impact we have had on it.

E. Rivers around the world have been significantly altered.
  1. Part of the year not a single drop of the Colorado River actually reaches the ocean because it all gets removed to grow food in places like California's Imperial Valley.
  2. Water resources across the United States have been diverted and rerouted to provide drinking water for its growing populations.

V. One of the largest effects of human activities has turned out to be on the atmosphere.

    A. Carbon dioxide and other greenhouse gases have been increasing in the atmosphere.

    B. Pollution from particulates and aerosols also causes an increase in cloud cover.

    C. Sulfur oxides and nitrous oxides have significantly increased the acidity of rain.

        1. Acid rains have removed life from many lakes and ponds that cannot handle the high level of acidity.

        2. Added acidity in the rain has increased chemical weathering, and therefore the rate at which our land erodes.

    D. Carbon dioxide levels had fluctuated within a narrow range for millions of years, but starting with the burning of coal during the Industrial Revolution about 200 years ago, the levels began to increase rapidly and are now on their way to levels they haven't been at since the Cenozoic era.

    E. Atmospheric pollutants can both heat and cool the land.

        1. Our increase in carbon dioxide and methane production leads to heating.

        2. Clearing the forests and the production of aerosols have both had cooling effects.

        3. Taking into account all factors, the net effect is an increase of heating of about 1.5 watts per square meter.

    F. There are increases in droughts, especially in the western United States (which has caused an increase in forest fires).

VI. All indications were that until the last 200 years, we were heading back into another period of Ice Ages. However, estimates are that within the next 100 years the temperatures will increase by 2–5 degrees, which is an enormous amount.

    A. In many parts of the world, temperature increases could be more than 8 degrees by the end of the 21st century, bringing droughts to many parts of the world that are already environmentally stressed.

    B. Glaciers are melting, sea levels are rising, and regions along the eastern United States are sinking from pumping water and petroleum out of the ground. These areas will be at tremendous risk of flooding.

    C. The situation for eastern Asia is even more critical because coastlines there contain some of the world's densest populations.

    D. Returning to the population levels of the 1930s would be ideal, but war, disease, and famine (the traditional ways of controlling population) are not very good options for doing this.

VII. One of the greatest impacts we have made is on the biosphere.

    A. Much of this is through our trade and travel all around the planet.

    B. Countless species have been introduced to new regions by human activities, either intentional or accidental. We have no idea what the environmental consequences of this transfer of organisms will be.

    C. A big part of this transfer is germs. Bringing germs into other parts of the world can have devastating effects on population.

VIII. The extinction of organisms is currently happening at a catastrophic rate due to pollution in the waters and atmosphere, changes in habitat and climate, and the transfer of organisms around the planet.

    A. Inadvertently, we have single-handedly brought about the end of the Cenozoic era. We have ended a 65-million-year era of time.

    B. Many of the mistakes we are making are not irrevocable. We need to understand and change them, not ignore or deny them.

    C. We are making mistakes that any civilization or species would do as it makes the leap from being just the latest of a long chain of simians and hominids to a species that can teach a course about it.

Recommended Reading:

Diamond, *Collapse.*

Pimm, *The World According to Pimm: A Scientist Audits the Earth.*

Questions to Consider:

1. What do you think the geologic consequence will be of the continued paving of America?

2. The population of the southwestern United States is booming, groundwater supplies in this region are dwindling, global heating is causing severe droughts in this area, and already the Colorado River often doesn't reach the sea. What do you see as a possible solution to this situation?

Lecture Forty-Five—Transcript
Humans—Dominating Geologic Change

Welcome. In this lecture, I want to focus on the current most powerful agent of geologic change, us, humans.

There's a poster I see on the walls of airports for the organization CARE, and it shows the dusty face of what looks to be an African woman, and the title on the poster reads, "I am powerful." And every time I look at it, I think, "You have no idea." There's nothing new about life altering the planet. This has happened regularly over the past 4 billion years. What is remarkable, however, is the speed with which one particular species, humans, is doing this. There are many previous examples of life significantly changing Earth's surface. Photosynthetic cyanobacteria, 2.5 billion years ago, removed the carbon dioxide from the atmosphere and emitted free oxygen, which gave us ozone, allowing life on continents. Ocean worms in marine sediments ended the Snowball Earth runaway greenhouse huge temperature swings by churning up the sediments and getting methane out of the sediments. Corals and shells created limestone that now cover most of the continents. Leafy trees became a massive store of carbon dioxide, tremendously buffering climates.

And of course, single-celled bacteria and archaea caused or catalyzed almost all geochemical reactions. There's almost no such thing as geology at the surface that doesn't involve life.

So, what's so unusual about humans? Well, we have managed to significantly alter our planet in just 200 years. Remember, humans are so young! Six percent of all human beings that have ever lived are alive today, and 20% of all adults that have ever lived (because in the past, many humans died during childbirth) on this planet are alive today.

We've done this through a scientific understanding of how the earth works, just the things that I've talked about in this course—natural resources, how to channel energy, how to incorporate physics and chemistry with the aspects of this earth—and through the development of this information (what we call engineering and industry).

I have to be careful when I talk about this subject, because when I go through this list of our effects on the planet, it too easily sounds like a litany of bad deeds and moralistic judgments. Pollution, the ozone hole, mass extinctions of animals, global heating, and on, and on, and on. It can get depressing, and nobody wants to hear it. But that's not what I'm saying here at all. Again, I'm a geologist. I take a long look at things. This is all inevitable. It's because we're an infant species. We're like a child, a big child, that doesn't yet know its powers. And because we're young, we make mistakes.

Each mistake we make, however, is an opportunity to teach, and the hope is that the lessons are learned when the mistakes are small so that the big mistakes don't happen. I mean, people don't intentionally destroy their homes, right? Humans have done that at times, but only because we're so young and just beginning to realize what we're doing. I mean, we've never been here before. It's inevitable that we're going to do some things wrong. Believe me, if there are other species in the galaxy that have reached this same stage of development, they're making plenty of mistakes, too.

I don't want to pretend that there aren't any mistakes; that's societal neurosis. You don't want to deny that they've happened; that's lying. What you want to do is learn from them while they're still small so that you don't make the big ones. There are plenty of examples of mistakes that were caught before they became larger. I talked about the ozone hole already. Ozone holes are a really bad idea. But, we passed legislation that stopped the production of chlorofluorocarbons around the globe, to a great degree, and the growth of the ozone hole has largely stopped and maybe even begun to reverse.

Leaded gasoline was a really bad idea. It washed off roads, into streams, and into people's water supplies. But in 1973, the U.S. began to phase out lead in gasoline, and lead levels in streams have dropped considerably and are almost gone. Atmospheric nuclear testing was a really bad idea. It led to clouds of irradiated isotopes that got into crops that got into milk. The testing moved underground, and now nuclear testing is almost nonexistent around the whole globe.

Now, the presence of countries makes things a little bit complicated, because lessons learned by one country can sometimes have to be relearned by other countries at a much later date. In one sense, the countries are like a bunch of children, all of different ages. Some have already made their mistakes and are learning from them, some are a little bit behind.

What I'm going to do here in this lecture is to list a few areas where humans are rapidly and powerfully altering their world. You can think of it as a report card, if you want, but no one should be sent to his or her room because of it. There's no shame involved here. We're just trying to figure out how to do this. However, some dinner table discussions might be a good idea.

Well, let me start at the top, literally. We're putting a whole lot of stuff up into space. It's an area you don't normally think about as having a large impact from humans, but there are currently more than 600,000 objects of at least a centimeter in size that orbit around the earth. These are remnants of previous satellite launches.

The United States Strategic Command actually tracks the locations, continuously, of the largest 10,000 of these, most of which are larger than a meter in size. They pose serious hazards to the international space station, any other missions out in space, and all the satellites that we have up there. Actually, one person so far—Lottie Williams, in Tulsa, Oklahoma on January 22, 1997—was actually hit in the shoulder by a small piece from a Delta II rocket fuel tank that had been launched the previous year.

Humans are significantly changing the appearance of many parts of the land, of course. Erosion rates have increased dramatically, primarily through the removal of vegetation and increased agriculture, but also a lot of other activities. For the past 10,000 years, humans have been removing forests in order to make lands for livestock and agriculture. That's nothing new. This, however, has had a tremendous effect on the biosphere, and currently—this is a remarkable statistic—about 35% of the world's land, that's one-third of all the continents that are not covered with ice, are now used for human foods, for growing crops and grazing livestock in order to feed us. And that number is rapidly rising. Each year—I mentioned previously when I talked about soils—about 70 gigatons of rare topsoil is lost each year. Stuff that takes thousands of years to create is being washed away every year, though that number is decreasing as better agricultural practices have been developed.

Highway road cuts have rapidly increased erosion in many parts of the world. Wetland regions have been largely drained or even filled in, removing the water from the land. This makes regions more arid and makes for much larger seasonal temperature changes, because you remove that water buffer. Remember that water has a very high heat capacity, latent heat. Taking those wetlands away from a particular region removes its ability to buffer its seasonal temperatures, and also makes the areas more arid. They have, therefore, less rain, which makes the area more arid again. Now, sometimes this process is inadvertent. An interesting example came at the end of the Little Ice Age, where it was so cold in Europe that European settlers in North America killed more than 50 million beavers in order to make fur coats and hats for Europeans. Well, removing the beavers removed the dammed up streams, and the New England wetlands disappeared. It actually has made winters in New England much harsher, because you've removed a huge portion of the surface water there. In the process, you've actually added to the increase in cold temperatures, because you've removed a significant amount of biomass from these regions.

The amount of the U.S. that is now paved is greater than the area of the state of Ohio and increasing rapidly. The paving affects land temperatures, which affects storm patterns. As I talked about previously, in streams it also affects water flow patterns, because it increases the mount of water that rushes into the streams and out into the oceans during storms. In other words, you prevent water which should be going into the ground to recharge our groundwater. You're taking that water and washing it out into the sea. In urban areas, that has become a huge problem.

Humans are currently generating waste at a staggering pace. There is about 450 million tons of garbage generated every year in the United States. That's about a ton per person per year. However, recycling has gone up considerably and quite dramatically. In 1991, recycling reclaimed, recycled, and reused about 11.5% of U.S. waste. That number went up to 32%, about one-third of waste, in 2001. The amount, of course, is much lower in other countries that aren't as far along on their recycling programs. China, for instance, dumps one-third of industrial waste and about two-thirds of human waste untreated, and China has a whole lot of people. Recycling is, in general, on the rise. There are other places, like Europe, however, which recycle even much higher percentages of their waste than in America. This is a lesson that's being learned.

Humans have had significant effects on streams and lakes. In America, government legislation has been extremely successful in reducing pollution. Streams are recovering well. The situations, however, are getting worse in other countries that are developing their industrial complexes. In America in the 1960s and 70s, many rivers and lakes were polluted to the point that they were undrinkable, unswimmable, [and] unfishable. They were unusable. Where I grew up in northern New Hampshire, there was a river there, the Androscoggin, that was so polluted that it filled the entire valley with an incredible stench. Well, that river is largely clean now.

A textbook example back at that time, however, was the Cuyahoga River in Cleveland, which had been filled with so many industrial chemicals and hydrocarbons that it actually caught fire and burned two different times, in 1952 and again in 1969. Well, a few years later the Clean Water Act went into effect, between 1972 and 1977, [and people] began cleaning up these lakes and streams, and it's really a tremendous success story. Life has returned. These streams are now mostly clean. Swimming and fishing has returned. The report card is not perfect; many streams in this country still have very high levels of pathogenic bacteria that are largely due to livestock and human wastes. Part of the reason for this is that our population is increasing so rapidly, and so are our food supplies.

Groundwater supplies in many parts of the world are significantly contaminated, as I talked about previously when I talked about the whole groundwater system. This comes from a variety of sources: industry, dumping, and petroleum spills. In farming areas, it's agricultural chemicals. Unlike streams, groundwater takes much longer to recover. It may be thousands or tens of thousands of years before the groundwater recovers from the impact we've already had on it.

The courses of many streams or large rivers are diverted or dammed. Remember the world's largest engineering project, the attempts to keep the Mississippi River in its current channel? Well, there are rivers around the world that have been significantly altered or used. One example is the Colorado River. Part of the year, not a single drop of the Colorado River actually makes it into the ocean. It's entirely withdrawn for human purposes—partly for drinking water, but mostly to grow our food, the crops in California's Imperial Valley. Remember that one-third of the world's surface that isn't ice is being used to feed us. In fact, water resources across the entire U.S. have been largely diverted and rerouted to places like Los Angeles to provide drinking water for the 13 million people there (Or is it 14...or 15 million?), the rapidly increasing populations in some of the cities.

One of the largest effects of human activities has, of course, turned out to be on the atmosphere. As I've alluded to several times during the lectures on climate change, carbon dioxide and other greenhouse gases have been increasing in the atmosphere, but it's not just greenhouse gases. Pollution comes from other areas, particulates and aerosols, often causing an increase in cloud cover. Sulfur oxides, nitrous oxides—these materials in the atmosphere have significantly increased the acidity of rain. Remember that I talked about rain having, normally, a pH of 5 to 6. That's a level of acidity. It's slightly acidic, normally. Well, there are rains now with a pH of vinegar, 2.4, or lower, in some cases. It's quite common for rains to have pHs of 4 or less. Acid rains have removed life from many lakes and ponds, life that cannot handle this level of acidity.

And interestingly, the added acidity in the rain has increased chemical weathering on our planet, the rate at which our land erodes. Remember my example of Cleopatra's Needle in Central Park? When you add this increased level of acidity in the rain, it increases your global erosion rates. We have actually greatly accelerated the rate at which the whole rock of the planet is eroding.

As I also previously mentioned in the lectures on climate, temperature and carbon dioxide have historically paralleled each other. Well, by the year 2007, the levels of carbon dioxide have increased up to 380 parts per million, which represents a 20% increase, a one-fifth increase during my lifetime alone. It's interesting: Carbon dioxide levels had been steady for centuries and millennia before that at about 280 parts per million, for a while. In fact, if you actually look at carbon dioxide levels going back over hundreds of thousands of years, the levels of carbon dioxide as a volume of the atmosphere had fluctuated between about 200 and 280 parts per million. Well, starting with the Industrial Revolution about 200 years ago, with the burning of coal, they began to increase rapidly. They are now well on their way to a level that they haven't been since maybe the Cenozoic—a long time ago.

The effects of humans on global warming are actually not so simple, because our atmospheric pollutants both heat and cool in places (as well as other effects that we have). Carbon dioxide and methane are greenhouse gases, so our increased production of these obviously leads to heating. Clearing the forests, however, which we've done, increases the surface albedo, the amount of sunlight that gets reflected and not absorbed. That's actually had a cooling effect on the planet.

The productions of aerosols into the atmosphere—these materials reflect sunlight and also prevent sunlight from being absorbed by the ground. That also has a cooling affect. When you take all the different factors, however, and bring them together, the net effect is, in general, an increase of heating of about 1.5 watts per meter squared. During the past century, the mean global temperatures have gone up about two-thirds of a degree Celsius, about 0.7 degrees. However, we're currently adding 7 gigatons of carbon to the atmosphere every year.

Now, this has caused many regional climate changes as well as a general global change. Remember that I talked about how the increase in temperature may be a global phenomenon, but the regional changes can be very, very variable. In general, there are increases in droughts, though some areas have actually received more rain. One result of having the increase in droughts, particularly in the western United States, is a tremendous increase in forest fires, which has been devastating, causing tremendous amounts of damage. Sea level rise is occurring, and there is perhaps even an increase in storms in the North Atlantic.

It's interesting. All indications were that until the last 200 years, we were heading back into another period of ice ages. This isn't surprising. This is what follows every interglacial period—remember, driven by those Milankovitch cycles. You have a few thousand years of warm temperatures, and then the temperatures gradually cool, and then for tens of thousands of years, they continue to cool until they rise very rapidly for another interglacial.

Well, global temperatures actually peaked about 7,000 to 8,000 years ago and have been continuously cooling over the past millennia. But those temperatures stopped cooling and began rising right at the end of the Little Ice Age in the mid-1800s, and they haven't stopped rising ever since.

By conservative estimates, global temperatures will probably rise by 2 degrees by the end of the 21st century. By more generous estimates, it may be 5 degrees, so the temperatures within the next 100 years or less will actually increase 2 to 5 degrees globally, which is an enormous amount.

However, what is even more significant is that if you look at a map of the predictions of where temperatures will rise most rapidly, you see a very alarming pattern. The largest temperature increases occur on land and not on oceans. Of course, water has a high heat capacity; it takes a lot of energy to raise the temperature of water. Land temperatures increase much more rapidly. So predictions are that the temperature increases on many parts of the land could be more than 8 degrees higher by the end of the 21st century, especially in regions like equatorial South America, or even very northern latitudes, northern North America and Eurasia. An 8 degree change in temperature would have significant changes on climate, bringing droughts to many parts of the world that are already environmentally stressed.

Another effect of an increase in temperatures, of course, is that glaciers are melting. Nearly all alpine glaciers have thinned over the past 4 years. In fact, the global mean alpine glaciers—the glaciers in mountains—the alpine glacial thickness has actually decreased every single year except for three since 1960, and the total amount of thinning of these glaciers is about 50 feet, about 15 meters. That rate, however, is rapidly increasing.

Of course, Greenland and Antarctica comprise most of the ice on the land, and they're melting, too—rapidly. In fact, some large blocks of the Antarctic ice sheets the sizes of states like Connecticut have begun to break off the continent and float off into the ocean and melt. Sea level has already risen 20 centimeters in the past century. By conservative guesses it would be another 20 centimeters by the end of the 21st century, but more likely and more generous guesstimates are on the order of 60 centimeters. That may not sound like a lot, because that's less than a meter, but that puts huge parts of the eastern United States seaboard at a very high risk from storm damage.

Not only is the sea rising, but storms are getting worse in the North Atlantic, and we have many regions along the eastern U.S. that are sinking from pumping out water and petroleum from the ground. It's not just Louisiana, and not just Florida, the places that you would expect. It's all along the coast, from Texas, to Alabama, and Georgia, the Carolinas, Virginia, Maryland, Delaware. All these areas have huge parts of the land that are within a meter or a few meters of sea level, and those will be at tremendous risk as the sea level continues to rise.

If you actually look at a similar map for eastern Asia, the situation is even more critical. Because we see that many of the areas of the world's densest populations—places like Bangladesh and India, coastal China, these populations on the order of hundreds of millions of people, in some cases—live right at sea level and are at a very high increase in risk of disaster from ocean storms.

Talking about people, the population of the human species is perhaps the single most important fact in this whole discussion. The number of people on the earth has more than doubled during my lifetime alone. The population was 3 billion when I was born. It's now approaching 7 billion, and we just don't know how many people our planet can reasonably hold.

There are some scientists who study the topic of human sustainability. In other words, what conditions would be required to have humans who peacefully exist on the planet for indefinite amounts of time? Some of their estimates are that the human populations couldn't exceed 1 to 2 billion in order to do that. Well, the population of the world hasn't been 2 billion since 1930. We would need to get back to population levels of that time. How do you get there? How would we do that? Through education? Through government control? Through war or disease or famine, which are the traditional ways of controlling population? None of those are very good options, but we can't keep expanding exponentially.

One of the greatest changes we've had on the earth is on the biosphere. Everything that I've talked about so far in this lecture not only affects the planet, it affects all the organisms that live in these regions, and not just people. Interestingly, one of the greatest impacts that humans have had on the planet, both geologically and in terms of the biosphere, is through our trade and travel all about the planet.

In the last couple hundred years, 97% of the species on Hawaii have been recently brought there by people visiting it. Whole environmental systems are reeling from the sudden takeover of regions by organisms that may have no natural predators.

Remember that I gave the example of the prickly pear cactus that was brought to Australia for a garden and almost took over the continent because it was so well suited to those dry climates and had no natural predators. It wasn't until the *Cactoblastis* moth was introduced, a natural predator from Central America, that the prickly pear population was reduced. Well, there are countless other examples of species that have been introduced by humans, either intentionally or accidentally, that have then spread explosively. Kudzu vines in the southeastern United States. Starling birds, which were brought over in the 1800s from England, now spread all across North America. You see these large enormous flocks of black birds, essentially pushing out most other bird populations in regions where they've settled. The African honeybee that came into South America. Lampreys and zebra mussels brought into the Great Lakes. The list of such sudden introductions is enormous, and we still have absolutely no idea what the environmental consequences of all this transfer of organisms around the planet will be.

Part of it—a big part of it—is germs. Before western settlers came to North and South America, it's estimated that 20 million Native Americans lived there. With the introduction of a variety of bacteria and viruses, within a short amount of time that population had crashed to about 1 million people, 5% of the population. The same thing happens not only with humans, but with other organisms as well. You bring germs rapidly into other parts of the world, and it can have a devastating effect on population.

I saved the biggest change for the end of the list. We don't know how many species of life there are on the planet; it's probably millions. Most are in the tropics. Most are probably still undiscovered. They range from tiny, simple, single-celled archaea and bacteria, to the enormous blue whale, with a whole heck of a lot of beetles in between.

The extinctions of organisms, we know, are a natural process. They've happened throughout history. But they're currently happening at a catastrophic rate due to the pollution we've put into the atmosphere, and water, and ground, and changes in habitat, the climate change, and the transfer of organisms around the planet. Estimates vary widely, but it's probably a conservative guess that 20% of life forms that are alive today will be extinct by the end of the 21st century. Have extinctions occurred at that rate before? Maybe—large meteorite impacts. But you know what happens to the geologic record when there are large periods of mass extinctions? These represent the boundaries between geologic time eras.

The Cenozoic era, that we're in now, began 65 million years ago when a large meteorite impact killed off the dinosaurs. In an instant, the course of the planet's history was changed, and we are doing the same thing. Inadvertently, accidentally, we have single-handedly brought about the end of the Cenozoic era. We have ended a 65 million year era of time in a very short amount of time. We have reached that level of geologic power.

Does this list of topics bring people down? It's my hope that it wakes them up. Some of the mistakes are irrevocable. I mean, once a species is extinct, it's extinct. But many of the mistakes we're making are not irrevocable, and we've already proven that many of the mistakes we've made can be changed. We need to understand them, not ignore or deny them, but change them. We need to be sure that we're not ashamed or embarrassed by them. I mean, when a child makes a mistake, do you embarrass it, or shame it, or punish it? No, you teach it. You let it learn.

We're in the same situation. We're making mistakes. For many of them, there's no shame in them. It's what any civilization or species would do as it makes that leap from being part of a long chain of, in our case, hominids and simians, to a species that can teach a course about it. We're the most powerful force on the planet. We can do this.

Lecture Forty-Six
History of Life—Complexity and Diversity

Scope:

Life on Earth began at least 3.85 billion years ago. The path of evolution since then has been a remarkable one, and an integral part of Earth's story. For the first few billion years, life existed as simple single-celled prokaryotes (bacteria and archaea). Sometime after 2 billion years ago more complex single-celled eukaryotes evolved. Multicellular life began less than a billion years ago, and only a half-billion years ago developed hard shells and skeletons. There are two ways to look at this. The path of evolution from worms to fish to amphibians to reptiles to early mammals to humans shows increasing degrees of complexity. However, on a tree of genetic diversity, all of multicellular life is a single small branch (variations on a single theme) while the real diversity occurs at the unicellular level, which probably happened early on. In fact, it is possible that no new metabolic strategies have evolved for billions of years. Even with multicellular life, an incredible diversity of life forms existed during the "Cambrian explosion" of life, but the plants and animals that survived it were mostly variations on a few successful biological strategies.

Outline

I. This lecture will discuss evolution and the history of life on Earth from the perspective of how conditions on a planet affect life.

   A. The evolution of life is intimately interconnected with the environmental conditions that shaped it.

   B. It cannot be emphasized strongly enough how powerful evolution is as a tool for solving the biological riddle: What kind of life is best suited to a particular environment?

   C. The simplicity of natural selection makes it powerful. The better model survives—gets the food, reproduces, and has offspring—the others don't.

II. Evolution on Earth has three major components to it: reproduction, mutation, and natural selection.

A. Whether it occurs asexually or sexually, reproduction allows the transmission of genetic code, DNA, from one organism to its offspring.

1. When some parts of the DNA are not copied correctly, genetic drift occurs, allowing species to change over time.

2. If a single population is spread out geographically, different parts of the population can drift genetically so that they can no longer interbreed. They become separate species.

B. Sometimes these genetic mutations are significant and allow the organism to better survive.

1. Earth plays a role in this mutation through solar radiation, which damages DNA structures.

2. There is an optimum mutation rate: If mutation is too slow, the species cannot keep up with climate change. If mutation is too high and there are too many birth defects, the species is unstable.

C. Natural selection occurs when an individual organism lives long enough to reproduce. As climate and geologic and geographic conditions change, species that may have been previously well-suited to their environments may no longer be so.

III. The path from single-celled organisms to humans is a remarkable story.

A. The only record we have of the first organism is the DNA of modern organisms, which we have been able to interpret and read back in time.

B. The first step was the creation of organic compounds, making the stuff of life. Amino acids are the building blocks of DNA and they form naturally in the conditions that existed in the earth's oceans.

C. The next thing you need is a cell membrane. Many amino acids naturally polymerize, and join in masses to form "peptides." Peptides formed "microspheres," and microspheres made of lipid bilayers (from fatty acids) provided the basis of cell membranes.

D. Next you need biochemistry and catalytic activity. The microspheres began to grow and to interact with both each other and the outside environment by ion exchange across membranes. Natural selection evolved because the molecules that replicated most efficiently began to multiply and take over.

E. The first life consisted of single-celled organisms. All of the biochemistry in multicellular animals, plants, and fungi also occurs in single cells.

F. Early life was in the form of single-celled organisms called prokaryotes, which still dominate the biosphere. They come in two major forms: archaea and bacteria.

   1. The first fossil evidence is from 3.85 billion years ago.

   2. Almost as soon as there were oceans, there was life on Earth.

   3. The early archaea and bacteria experimented with a tremendous number of metabolisms and biochemistries.

   4. There was a huge amount of organic material to feed on, so organisms quickly evolved to eat methane, nitrogen, sulfur, and many other different materials—there was a great amount of biochemical creativity.

G. About 2.5 billion years ago, one form of bacteria, cyanobacteria, developed photosynthesis. The bacteria figured out how to make its own food.

H. Simple single-celled organisms, sometime between 2.1 and 1.2 billion years ago, evolved into more complex single-celled organisms we call eukaryotes.

   1. Eukaryotes developed a symbiotic relationship with mitochondria and that became the foundation for all multicellular life, which evolved from eukaryotes.

   2. Eukaryotic plants developed a symbiotic relationship with another protobacterium called a chloroplast, and they became the basis for all modern plants.

I. There are three major life forms.

   1. Bacteria and archaea are separate branches, although they have a common ancestor from before 3.8 billion years ago.

   2. The third branch, eukarya, also consists of single-celled organisms.

3. Animals, plants, and fungi are a small branch off of eukarya.

IV. The most significant period of evolutionary diversity for multicellular life began about 600 million years ago, the start of the Cambrian period.

   A. Starting 540 million years ago, shells, spines, claw, jaws, and all sorts of other creative body parts became common. The most important fossil outcrop from this time is the Burgess Shale in Alberta, Canada.

   B. Life started on a long road of experimentation, with life evolving to fill every possible ecological niche where water existed.

   C. Reproduction, body metabolisms, and shape and style of organisms changed to survive in different environments.

   D. Animals learned to move through a diversity of mechanisms.

   E. Plants and fungi figured out how to disperse their spores and seeds.

   F. Animals learned to find food in a variety of ways.

   G. As fast as predators evolved, prey evolved a remarkable array of techniques to avoid being eaten.

   H. Some of the most fascinating evolutionary survival mechanisms involve symbiotic relationships between different organisms.

      1. Sometimes a symbiotic relationship can look like a single organism.

      2. Sometimes symbiosis occurs between separate cooperative individuals.

   I. A whole host of mechanisms for survival have evolved.

V. The path that led to humans is remarkable.

   A. Perhaps the most important organism, from a human perspective, that we have found from the Burgess Shale is a flatworm called pikaia, because it is the first known animal with a primitive spinal cord.

   B. The evolutionary path from flatworms to mammals only took 300 million years because not much change was needed. All vertebrates are very closely related.

C. The structure of a vertebrate is very efficient and effective and nearly all large organisms are vertebrates.

D. Flatworms evolved into ray-finned fishes and then lobe-finned fishes.

E. Lobe-finned fishes had sturdy fins that let them go on land to find food.

F. Amphibians evolved from fish and dominated the shorelines from about 350 to 300 millions years ago until reptiles evolved from them and eventually replaced them.

G. Reptiles diversified into a huge variety of forms. The path to mammals is led through a reptile called a therapsid.

VI. The fossil record is filled with large extinction events that mark out the many divisions between the geologic periods.

A. The largest of these occurred at the end of the Permian, 250 million years ago. About 90% of all marine species died off. On land, 70% of all vertebrates went extinct, including almost all of the reptiles.

B. The loss of most of the large reptiles allowed for the dominance of the dinosaurs and the start of the early mammals.

C. Mammals weren't really significant for more than a hundred million years; they simply couldn't compete with the dinosaurs.

D. The extinction of the dinosaurs didn't immediately cause the dominance of the mammals, but opened the door for their expansion and diversification.

E. The Cretaceous/Tertiary extinction 65 million years ago was caused by the impact of a large meteoroid that hit the Yucatan Peninsula and created the Chicxulub crater.

F. The earliest primates existed before this meteorite impact and were most closely related to flying lemurs and tree shrews.

G. Apes and monkeys diverged from them about 30 million years ago, and humans diverged from other apes about 13–10 million years ago.

H. This history is not an ascent of man, but rather a random walk of evolution, shaped by unique climate histories that existed on Earth at the time. It will likely not end with the species *Homo sapiens*.

Recommended Reading:

Schopf, *Cradle of Life*.

Westbroek, *Life as a Geological Force*.

Questions to Consider:

1. What do you think it is about the vertebrate structure that makes it so successful as a large animal form?

2. What kind of new *Homo* species will out-compete humans? By what criterion will it be evolutionarily selected?

Lecture Forty-Six—Transcript
History of Life—Complexity and Diversity

Welcome. In this lecture, I want to talk about evolution and the history of life on Earth. Now I think you'd agree that life, the biosphere, plays a vital role in how the earth works and the whole story of our planet, because frequently during this course, I've talked about how life affects Earth's surface, the importance of the biosphere for geology. In previous lectures, I've talked about how climate has affected human history. In fact, in the last lecture, I talked about how humans are currently shaping all of Earth's history. In this lecture, I'm going to step back and look at the broader process of having life on a planet from the perspective of how conditions on a planet affect life. How does life even begin on a planet? Why does it change? Why do some species evolve and others become extinct? Then I'm going to give an overview of how we came about, the path that led from one-celled organisms to humans.

Now, it may seem a little bit out of place to talk about the history of life in a geology course, but the evolution of life is intimately interconnected with the whole environmental conditions that shaped it. The biosphere is a vital part of Earth science. But we have the fossils. Paleontology has been a vital part of geology since its very beginning.

I can't emphasize strongly enough how powerful evolution is as a tool for solving this biological riddle: What kind of life is best suited to a particular environment? I mean, years of computer and wind tunnel testing, and we still can't come up with a more aerodynamic model for flying than a hawk! With light, hollow bones and feathers, it's an amazing structure! We can't come up with a better hydrodynamic model for swimming than a shark, with its skin and fins and gills. It's incredible!

But the answer, the reason that occurs, is quite simple; if there were a better way, it would exist. It would survive. It would get the food, reproduce, and have offspring, and the less efficient models wouldn't. It's the simplicity of natural selection that makes it so powerful. The better model survives, the others don't. Over time what you end up with is what works best.

But I'm getting a little bit ahead of myself. Let me define evolution. Evolution on Earth has three major components to it: reproduction, mutation, and natural selection. Now, reproduction is an amazing process, but I'm going to leave that to biologists to explain. Whether it occurs sexually or asexually, the point is that you transmit the genetic code, the deoxyribonucleic acid (DNA), from one organism on to offspring. It's that continued transmission of genetic information, that's what reproduction does.

However, that reproduction does not occur perfectly. There are mistakes, and that's really important. In fact, this is where Earth's environment begins to play a role, because when that genetic material gets copied, sometimes the parts of the DNA aren't copied correctly. And what that does is it allows for genetic drift. Over time, your species will change. Now, it can happen slowly. If you have a single population that's spread out, separated geographically, over time different parts of that population can drift genetically so that they can no longer interbreed. They actually become separate species. The classic example is Darwin's finches on the different islands of the Galapagos Islands. Separated geographically, they evolved into separate organisms. They follow different evolutionary paths.

Sometimes those mutations, however, are big, significant changes. Most of these we call birth defects. They're bad for the organism, and they kill it. But sometimes you get a genetic change that's beneficial. It's an improvement; it allows the organism to survive better, and you end up with a new species. The biologist Steven J. Gould called this "punctuated equilibrium."

The way the earth plays a role in this is that mutation is partly affected by solar radiation, which damages DNA structures. Remember that before ozone, before 2.5 billion years ago, the sun's radiation at the surface of the continents was too intense for life to exist. There was no life on continents. The mutation rates would have been too high. Life had to begin in the oceans.

In fact, there turns out to be an optimum mutation rate. If your species changes too slowly, if your mutation rates are too low, you are unable to keep up with climate change. You just don't survive. However, if your mutation rates are too high, you get too many birth defects, and your species is unstable.

The third part of evolution is geology's big contribution, and that's natural selection. Natural selection occurs when an individual organism lives long enough to reproduce and pass its genetic information on to its offspring. It doesn't happen at a species level; it happens at an individual level, and it's a form of immortality. Your genes live on in the following generations. Organisms that are well-suited to particular conditions at that point in space and time survive and reproduce, and those that are not well-suited die and do not reproduce. And here's the key: As climate, and geologic, and geographic conditions change, species that may have previously been well-suited may no longer be so. They get replaced by different species. As we saw repeatedly, the climate on Earth's surface is always changing, and all the different styles of reproduction, and transpiration, and eating, and avoiding being eaten, etc., have all come about through natural selection in particular environments that are constantly changing.

Again, I'm getting ahead of myself a little bit, because I haven't said yet how we ended up with multicellular life—people! It turns out that the path from single-celled organisms to humans is just a remarkable story, and I'm going to just be able to summarize it in a few short descriptions, but think of it as a *Homo sapiens* family tree, so to speak.

We don't know exactly how life began on Earth. We weren't there, and there are no fossils. There are no organisms remaining from the first times that life existed on this planet. We don't know what those first organisms were, but we do have a record of a sort. We have DNA. Our DNA and the DNA of all other organisms on this planet retain a record that goes back to that first common ancestor. And we've actually been able to interpret the DNA and read back in time to get a sense of what happened.

To briefly describe the process, the important thing is that life originates naturally and relatively quickly. In other words, if you start with the elements and molecules that existed on the surface of the earth, you end up with life. It's inevitable. You take that same situation again, and you might get life occurring again, because all the building blocks for life were here, in place, and available. And they were in a large, liquid ocean that provided that aqueous solution within which all the biochemistry could take place.

Now, the first step towards creating life from what was initially inorganic material was the creation of organic compounds, the basic elements that life has—making the stuff of life. There have been lots of experiments that have shown that if you start with Earth's early conditions—water, carbon dioxide, hydrogen sulfide, phosphate, ammonia, methane, all present in Earth's early atmosphere—and you take that material and heat it, or zap it with ultraviolet radiation, or freeze it, you get amino acids. Amino acids are the building blocks of DNA. For example, if you take hydrogen, and carbon, and nitrogen together, that makes a molecule of cyanide. If you take a whole lot of cyanide and expose it to ultraviolet radiation, like from the sun, it naturally bonds together to form the amino acid adenine. All the amino acids form naturally in the conditions that existed in Earth's oceans.

The next thing you need for life is a cell membrane. Well, it turns out that many amino acids, when you heat them, naturally polymerize. They join together in large masses to form long chains called "peptides," and sometimes these peptides will naturally form spherical shells called "microspheres." Now, the microspheres are not alive, but they are cellular in appearance. They actually absorb additional peptides on the inside of their structure, and they grow and divide; they add peptides to their shell. When they get too big, smaller pieces actually bud off to form new microspheres. There are lots of different microspheres, but the ones made of lipid bilayers, made from fatty acids, provided the basis of living cell membranes.

The next thing you need for life is the development of biochemistry, what we call "catalytic activity." It turns out that microspheres provided safe environments for those organic molecules to exist. An organic molecule is just a molecule that's carbon based. These materials began to grow, and interact with each other, and even react with the outside environment by ion exchange across the membranes. You had amino acids forming longer chains; nucleotides, forming RNA molecules; ribonucleic acid, which eventually could actually begin to split and self-replicate. At that point, you led to a process of natural selection, because the molecules themselves would begin to compete with each other inside the spherules. The molecules that replicated most efficiently began to take over and you had a situation that evolved very quickly toward the development of the most efficient biochemistry, which is what we have today.

Modern proteins form most of our biochemistry. They out-competed simpler RNA molecules, and they became the dominant compounds for all living cells. It happens naturally. If it were to happen again, it would happen the same way. Now, what's interesting is if it happens on another planet under different conditions, does it also happen the same way? Do you still get proteins and RNA and DNA? I don't know. That's what NASA and the space program are currently trying to find out.

The first life consisted of single-celled organisms. These microspherules began to carry out the organic chemical reactions that we call life. The story of the evolution of life from that point onward really has three parts to it. From a big picture perspective, it's very interesting. First of all, our biochemistry all works in one way. It was like that in the beginning, it will always be like that, and it's a direct result of the laws of physics, and chemistry, and the conditions on this planet. I mean, you could teach a course on the history of life and spend almost the whole course, maybe except for the last week, on single-celled organisms, because all of the biochemistry in animals, and plants, and fungi, and all that, all the proteins, and catalysts, and chemical reactions, occur in single cells. They haven't changed.

Life began with an incredible diversity at that single cell level, but not much complexity. They were all single cells. When multicellular life first evolved, life began to become much more complex, and there was a burst of diversity, all kinds of creatures. However, over time, life as we see it today, currently, on the planet is actually a refinement of just the most successful organism types. In fact, in a sense, it's actually less diverse. Let me explain. The initial life was in the form of these single-celled organisms. We call them prokaryotes. They're simple cells; they have no cell nuclei, or mitochondria, or other organelles. The prokaryotes took two major forms, archaea and bacteria, which evolved from some common ancestor.

The first actual fossil evidence of these goes back to 3.85 billion years, though geochemistry and rocks from that time suggested that life existed even before 4 billion years ago. That's amazing. Almost as soon as there were oceans, there was life on Earth. The earliest fossils are either in the form of stromatolites, which are mound-like structures that we think were colonies of bacteria, we see them off the coast of modern-day Australia today, or strands, algae-like strands of fossils. These early archaea and bacteria experimented with a tremendous number of different and very unusual metabolisms and biochemistries. There was a huge amount of organic material to feed on, so organisms developed to eat methane, and nitrogen, and sulfur, and many other different materials—a tremendous amount of biochemical creativity.

However, about 2.5 billion years ago, one particular form of bacteria, cyanobacteria, developed photosynthesis. It was a really hard trick to learn, but they figured out how to make their own food, and when they did, they radically changed the whole scene of life on Earth. Remember, that's the time that oxygen and ozone were created into the atmosphere. Sometime between 2.1 and 1.2 billion years—we don't know, we don't have good enough fossils to determine exactly the point in time, but we know that those simple single-celled organisms evolved into more complex single-celled organisms that we call eukaryotes. These developed a symbiotic relationship with mitochondria.

The mitochondria are the foundation of cells, of creatures of multi-cellular life. They initially existed as a separate form of bacteria. Mitochondria provide energy to cells. They have their own DNA, and they reproduce independently of us. At some point, mitochondria entered into another kind of bacteria and survived, to the benefit of both. They developed a symbiotic relationship, and this became the foundation for all multicellular life which evolved from eukaryotes. Eukaryotic plants actually developed a further complexity. They also developed a symbiotic relationship with another protobacteria called a chloroplast, which also has its own DNA, and they became the basis for all modern plants.

Multicellular life developed sometime after a billion years ago. And it's very interesting: If you look at the tree of life, what you see is that there are three major life forms. For instance, modern trees that show the diversity of life are done using the genetic changes, and if you use those genetic changes as a basis, you can see that bacteria and archaea are separate branches, but they have some common ancestor back beyond 3.8 billion years ago. But the third branch, eukarya, also consists of single-celled organisms. So, the whole tree of life is all single-celled organisms, all the diversity of different types of metabolisms and biochemistries. So, where are the animals, and plants, and fungi? They are a small branch off the corner of the tree of eukarya. They are slight variations. They're in there with giardia, and paramecium, and amoeba. All of the single-celled organisms that are more complex are very closely related to us.

The most significant period of evolutionary diversity for multi-cellular life began, however, not at a billion years ago, but some time about 600 million years ago, the start of the Cambrian period. The earliest multi-cellular fossils began at about 600 million years ago. They are fossils found in the Ediacara Hills of Australia. These fossils show the early ancestors of mollusks, and worms, and jellyfish. These fossils are very rare, because there are no shells or bones from this time. Starting with the Cambrian explosion 540 million years ago, we see the beginnings of shells, and spines, and claws, and jaws, and all sorts of other creative body parts.

The most important fossil outcrop from this time comes from a place in Alberta, Canada called the Burgess Shale, and we see the origin of predation and evasion. The act of natural selection showing up in the fossil record, evidence of claws and jaws for predators, and spines, and shells, and other hard body parts for prey. The Burgess Shale fossils display an unbelievably bizarre assortment of life. Creatures like the marella, and the opabinia, and the hallucigenia have no equivalent today. We don't even know how to classify them; they don't fit into any sort of life structure we see existing on our planet today.

Life at that time was experimenting incredibly creatively with cellular organisms. I mean, there wasn't that much competition initially, so anything that rolled off the assembly line would fill some ecological niche and could survive.

Though the number of species has increased in time since then, it's really only been a fine honing, however, of the few life types that are most efficient. Diversity of life at the start of the Cambrian was incredibly [high], with many approaches on life. Most got out-competed. They just weren't that efficient, and they didn't survive. Life has definitely become more complex, but it's been a refining of a few efficient mechanisms.

From the start of the Cambrian onward, natural selection started on a long road of experimentation with life evolving to fill every possible ecological niche where liquid water existed—on the land, underground, up in the air. Reproduction changed in many different ways—eggs, or spores, or, eventually, live young. Body metabolisms changed as did the shape and style of organisms in order to survive in different environments like deserts and jungles, fresh and salty water, deep ocean, and alpine tundras.

Animals learned how to move through an incredible, remarkable diversity of mechanisms: waving little flagellae, or running, flying, swimming, crawling—all these different mechanisms. Even plants figured out how to move. Plants and fungi figured out how to disperse their spores and seeds through fruit eaten by animals, burrs that stuck on animals, even exploding seeds. Animals find food currently with all five of their different senses and obtain it by burrowing, and grazing, sucking, chewing, poison. Even plants, like the Venus flytrap and the poly pitcher, have tried their hand at predation. They catch living organisms, insects.

But, as fast as predators evolve, so does prey evolve in order to avoid being eaten. And prey has evolved a remarkable array of techniques to survive: camouflage, evasion, deception, intimidation, confusion, armored protection. One of my favorites is an insect, the 17 year cicada, which stays underground for 17 years, which is a prime number so that no predator from its previous emergence 17 years ago could at all be prepared for the next large emergence.

Parasites have evolved, getting their food from some other organism. Your body is filled with them. Most of them are harmless. It's actually bad for business for a parasite to kill its host, because it wouldn't have any more food. Tapeworms can grow 50 feet inside your digestive tract, but they rarely kill you.

Some of the most fascinating evolutionary survival mechanisms involve symbiotic relationships between different organisms. I already mentioned, in previous lectures, the symbiosis between the tropical acacia trees that develop these large hollow horns that provide homes for whole ant colonies. It's incredibly common. Sometimes a symbiotic relationship can look like a single organism, like us—our bodies and the mitochondria that provide power in our cells—or lichen, which is a symbiotic relationship between a fungus and something that carries out photosynthesis. It can either be algae or cyanobacteria. Sometimes, however, symbiosis occurs between separate cooperative individuals. For instance, there's a Malaysian tree called the chempedak, which is pollinated by insects, gall midges, but the insects will only pollinate the tree if the tree gets attacked by a particular fungus. So, what you have is a symbiosis between a plant, an animal, and a fungus, and it benefits all three.

A whole host of mechanisms for survival have evolved. Some of them are even very bizarre. The lodgepole pine tree has cones that don't even open up until they are heated by a forest fire. I mean, that's designed not only to survive forest fires, but to be the first organism to take advantage of the fertile soil left over from the ashes. One of my favorites is the whole process of surviving the seasons. Think about it. If you're a plant with fragile tissue, soaking up water, what do you do about winter? How do you deal with ice and cold and snow? How do you deal with a planet that has a tilted axis and, therefore, has seasons? Well, it's not a big deal in the tropics, because it's warm there year round, but what about land far away from the equator near the North and South Pole? Well, one way is the evolution of firs and pine trees. Needles are long and skinny; they have lower surface area to volume ratios than leaves, and they survive through winter. Or you drop your leaves; you close up shop for the winter. It took more than 100 million years to figure this out, but as soon as it did, deciduous trees spread all about the continents. For land plants in general, it was just a remarkable transition to figure out how to survive on continents out of the ocean—the development of whole vascular systems, holding water inside of walls.

Let me get back to humans. I led off with the Cambrian explosion and the Burgess Shale fossils. Perhaps the most important organism that we've found from that fossil find, from a human perspective, is a flatworm called pikaia, because it's the first known chordate, an animal with some primitive spinal cord. Pikaia may well be the ancestor of all vertebrates, including humans. The evolutionary path from flatworms like pikaia to mammals actually only took 300 million years, because not very much change was needed. All vertebrates are very closely related. Remember that you share 50% of your genes with a mushroom. Well, you share 95% of your genes with a mouse; you have nearly identical body structures. You have appendages coming off a central spine, and for the appendages, you pick—arms, fins, wings—they're all structured similarly. The whole structure of a vertebrate is very efficient and effective, and nearly all animals are vertebrates. Natural selection favors vertebrates.

From flatworms onward was a fairly straightforward evolutionary route with some tinkering here and there. From 4 billion years ago to flatworms was a huge, enormous amount of time, but from flatworms to modern humans just took a few hundred million years. Well, flatworms evolved into early fish. The fish initially were jawless—this was back in the Ordovician period, 450 million years ago—things like *Agnatha*, which is an ancestor of the modern lamprey. These early jawless fish then led into jawed fish, like placoderms and acanthoderms. Some of them had massive, large, bony jaws, but no teeth yet. Interestingly, sharks and other fishes that have cartilage and not bone diverged off at this point.

From those early jawed fish came ray-finned fishes like most modern fishes, and then lungfishes, and then lobe-finned fishes like the famous living fossil, the coelacanth, *Latimeria*, that was found at the beginning of the 20th century off the coast of Madagascar. These lobe-finned fishes had sturdy fins that actually let them go up a little bit onto land to get the food, plants and insects, that were beginning to proliferate along shorelines.

Amphibians evolved from fish. That was pretty tricky and took a while, because it's hard to survive up on land. In fact, a very famous fossil, the *Tiktaalik*, was found in 2006. It's a type of organism called a tetrapod that's really a transition, somewhere in between fish and amphibian. It lived 375 million years ago, and from it evolved the early amphibians, things like *Acanthostega* and *Icthyostega,* that lived both in the sea and on land, but took advantage of the growing plant life on continents. Amphibians ruled the whole shoreline and near shore areas from about 350 to 300 million years ago, but reptiles evolved from them at about 300 million years ago and eventually largely replaced them.

Reptiles diversified into a huge variety of forms that walked, and slithered, and flew, and swam. Most of those early reptiles are all extinct, but some paths have led to our modern-day reptiles. One group of the reptiles, the thecodonts, evolved into crocodiles, the flying pterosaurs, and then dinosaurs, which then evolved into our modern-day birds.

The path to mammals led through a particular type of reptile called a therapsid. It was a very early reptile, closely related to the fan-back *Pelycosaur*. Therapsids like *Diictodon* and *Cynognathus* were warm-blooded. They had skin glands instead of scales, and mammals evolved from them about 200 million years ago. Now, for a while they were a very minor form of life. But, over time, they began, finally, to dominate.

It's interesting that the whole fossil record is filled with large extinction events. They mark out the many divisions between the geologic periods. The largest of these occurred at the end of the Permian, 250 million years ago, and it was a devastating extinction. It was worse in oceans than on land. In fact, 90% of all marine species died off, and almost all organisms themselves. It was the end of the trilobites, all the primitive fishes. On land, 70% of all land vertebrates went extinct, including almost all the reptiles.

It's not clear what actually caused the Permian/Triassic extinction. It's likely that there were several factors involved: severe volcanism, a meteorite impact, rapid climate change including glaciation, maybe even a loss of habitats due to the close of the supercontinent Pangaea. But the loss of most of those large reptiles allowed for the dominance of the dinosaurs and the start of the early mammals.

Mammals weren't really significant for more than 100 million years; they simply couldn't compete with the dinosaurs. The dinosaurs dominated almost all niches for large animals. The first mammals, like *Morganucodon*, were small, nocturnal, shrew-like animals. They had leathery eggs, but they suckled their young. Now, they eventually gave rise to the three forms of mammals that we see today: monotremes, marsupials and placentals. Monotremes, like the platypus, are mammals that are most like the early ancestors; they still lay eggs. Marsupials, like kangaroos and opossums, are also a more primitive branch of mammals. But most modern mammals are placentals, and they went through two large bursts of diversification about 85 and 55 million years ago.

Now, I've talked about the extinction of the dinosaurs 65 million years ago. Interestingly, that didn't immediately cause the dominance of the mammals. But the loss of the dinosaurs opened the door for the later expansion and diversification of mammals. The Cretaceous/Tertiary extinction 65 million years ago didn't just kill off dinosaurs; it killed off most other large reptiles and all sorts of other species of birds, and marsupials, and insects, and many of the marine animals like ammonites. The cause of the mass extinction, however, we do know; it was the impact of a large meteoroid that hit the Yucatan Peninsula, Mexico, and created the Chicxulub crater.

The earliest primates had existed before that meteorite impact. The earliest primates actually were most closely related to flying lemurs and tree shrews. Apes and monkeys diverged from them about 30 million years ago. Humans diverged from other apes about 13 to 10 million years ago. We have a very rich fossil record from about 10 million years onward of the steady evolution of humans—*Australopithecus*, which evolved about 4.5 million years ago. The *Homo* species began about 2.5 million years ago. It turns out that there were actually more than a dozen different identified species of *Homo*: *Homo habilis*, 2.5 million years ago; *Homo ergaster*, 1.9 million years ago; *Homo erectus*, 1.25 million years ago; the Heidelberg man, 600,000 years ago; *Neanderthals*, 230,000 years ago; and finally, 195,000 years ago, us, *Homo sapiens*.

It's important not to look at this history as an ascent of man, because it's nothing of the sort. It's a random walk of evolution that led to mostly dead ends. It's entirely shaped by the unique climate histories that existed on Earth at the time—impacts, ice ages, floods, volcanoes—and it will likely not end with the human species, *Homo sapiens*, as well.

Remember that all multicellular life was just a tiny branch off the eukaryotes' tree. It's variations on a single theme, but a very successful theme. In the next two lectures, I'm going to take this show on the road. I'm going to look for any other places in the solar system and the galaxy where life might exist to see if there are any other planets like Earth out there.

Lecture Forty-Seven
The Solar System—Earth's Neighborhood

Scope:

One of the best ways to understand how Earth works is to compare it to the other objects that formed along with it. Each of the planets and their moons has a very unique structure and history, which is strongly dependent upon its location in the solar system. The compositions of objects are a function of the temperatures that existed at the start of the solar system: rock and metal for the inner planets, ices for the outermost objects. The inner terrestrial planets, like Earth, are mostly rock and metal. The further out you go in the solar system, the less rock is found and the more that planetary bodies are comprised of ices and gasses. The Jovian planets grew large enough that they could hold on to hydrogen and helium, so they grew very large. Many of Earth's geologic processes (volcanism, earthquakes, erosion, etc.) are found on other planets or their moons. Examining this diversity of planetary bodies gives us a better appreciation of how unusual and unique our own planet is.

Outline

I. As the previous 46 lectures have shown, many remarkable and complex factors have gone into forming our planet and its biosphere.

   A. The last two lectures will review the many geologic processes that have not only created our planet, but continue to sustain it.

   B. This will be presented in the context of answering a simple question: Are there any more planets like Earth?

   C. We can initially start with our own solar system. Life exists here on Earth, so we know that our sun is a suitable star to foster the evolution of life.

   D. What are the chances of finding life on other planets like ours? If you start with Mercury and work your way outward across the solar system, you will notice some patterns emerging:

1. Each world in our solar system is very different from the others.
2. In spite of this, many share similar aspects of geology such as sand dunes, ice caps, volcanoes, rivers, and hurricanes.

II. The four inner planets of our solar system—Mercury, Venus, Earth, and Mars—are called the "terrestrial planets."

   A. These four are all similar to each other, but very dissimilar to the other four planets, the gas giants Jupiter, Saturn, Uranus, and Neptune.

   B. Like all of the terrestrial planets, Mercury has an iron core, a rocky mantle, and a crust. However, being closest to the sun, it has the highest percentage of iron, with a very large core.

      1. Part of Mercury's outer core is likely to be still liquid, as it has a strong magnetic field for its size.
      2. Mercury is very small, and so has a weak acceleration of gravity and no atmosphere.
      3. Mercury is geologically dead; its interior and surface have been unchanged for more than 4 billion years. This has occurred because of its small size: smaller objects have greater surface area-to-volume ratios, which causes them to cool off more quickly.
      4. Being close to the sun, Mercury is hot and has no water and, therefore, no life. Its temperature varies dramatically from -175°C on the side away from the sun to 425°C on the side facing the sun. Humans will never live there.
      5. Mercury has an unusual orbital pattern, rotating three times for every two revolutions it makes around the sun. This pattern of having an integer variation between revolutions and rotations is common in the solar system, and is a result of tidal resonance.

   C. Venus is close to Earth in size with a radius that is 95% that of Earth.

      1. Venus has a thick atmosphere that has a pressure of 90 atmospheres at the surface (90 times that of Earth's atmosphere) and is 95% carbon dioxide. Venus has a runaway greenhouse.

2. Venus has an iron core and probably a liquid iron outer core, but no magnetic field. This is probably because Venus rotates so slowly: It makes one counter-rotation every 243 Earth days.

3. The thick carbon dioxide atmosphere makes the planet very hot, with a mean surface temperature of 460°C, so there is no water on the planet (though there are traces of sulfuric acid in the atmosphere).

4. All these factors combine to make Venus an incredibly inhospitable planet for the existence of life and, as a result, NASA and other space programs have little interest in further exploration there.

D. Earth's moon formed at the same time Earth formed and has a similar composition, though with much less iron and therefore a much smaller core.

1. Because the moon is much smaller than Earth, its surface gravity is only one-sixth of ours, and it cannot hold on to an atmosphere.

2. Like Mercury, the surface of the moon has been largely unchanged for billions of years. Light-colored areas on the moon are highlands, which occupy most of the planet's surface. Rocks there date back to about 4 billion years ago. Dark areas on the moon formed more recently—about 3.5–3.0 billion years ago.

3. The dark areas are called "maria," which is Latin for oceans. However, the moon's maria are flat flood basalts, and no water has been found there.

4. The moon is the only planetary body we have visited, but without water, it is unlikely that it will ever support human colonies: It would be too difficult to transport water there.

E. Mars is the most Earthlike planet and the focus of a lot of attention because it has had ample water at its surface in its past and therefore may have once supported life.

1. With its thin atmosphere, Mars is roughly as cold as Antarctica, so it's almost livable. There are several Earthlike environments on Mars such as ice caps, deserts, sand dunes, and dust storms.

2. Mars has a thin atmosphere that is mostly carbon dioxide (good for plants!). The temperatures at the poles are cold enough that carbon dioxide exists there as small frozen polar ice caps.

3. Mars had a tremendous amount of volcanic activity about 3.5–3.0 billion years ago (like the moon), though some lava flows are much more recent. One of its volcanoes, Olympus Mons, is the largest volcano in the solar system.

4. Although there is no liquid water at the surface now, there is evidence that it may be frozen beneath the surface and even occasionally leaking out of cliff walls, and the NASA program is investigating this. If water is discovered on Mars, the planet will be the most likely site for human colonization.

F. Between the orbits of Mars and Jupiter is the asteroid belt, consisting of countless small rocky objects.

1. Of the million or so of these asteroids, 160,000 have been named. The largest of these is Ceres, which is now considered to be a dwarf planet.

2. The asteroids are likely the remnants of the early solar system, planetesimals that never became planets because of the strong gravitational force of Jupiter.

3. Asteroids have no water, no life, and haven't much interest for us as places to live.

III. The giant planets Jupiter, Saturn, Uranus, and Neptune formed with rock and metal like the terrestrial planets, but because they formed beyond the condensation point for ices, they also accumulated water, ammonia, and methane. This allowed them to quickly become very large.

A. Jupiter and Saturn are quite similar to each other, as are Uranus and Neptune.

1. Jupiter and Saturn are made mostly out of liquid hydrogen and helium, so the term "gas giant" is really a misnomer. They do not have solid surfaces, although they do have active weather in their atmospheres, where methane and ammonia form colorful bands that change as they move.

2. Both Jupiter and Saturn rotate very quickly: Jupiter's day is only 10 hours long, and because it is so large, the velocity of its rotation at the equator is about 45,000 kilometers per hour. This causes a very large Coriolis effect and the formation of many bands of atmospheric circulation.

3. Jupiter is the largest planet, with a diameter about 11 times that of Earth (and therefore a volume about 1000 times larger!). The large red spot on its surface is a storm that has been there as long as we've been observing the planet: at least several centuries.

4. Saturn has the lowest density of the planets, less than a gram per cubic centimeter. While all four of these planets have rings, Saturn's are the most prominent, though the thickness of the rings is remarkably thin: only tens of meters thick.

5. Uranus has an unusual orientation relative to its orbit, with an axis of rotation that is tilted on its side.

6. Neptune is 30 astronomical units from the sun, the farthest planet. It takes 165 Earth years for Neptune to make one revolution around the sun. Because of its tilted axis, it has seasons.

7. Neptune had a large storm spot that was observed in 1989 by the Voyager 2 mission, but was gone by 1994.

B. In our search for other Earthlike planets, there isn't much to recommend of the four giant planets. Their rocky surfaces are thousands of kilometers beneath thick mantles and atmospheres of ices and gases at unbelievable pressures.

IV. It is the moons of the giant planets, however, that have the more Earthlike geology, and even hold the most hope for the current existence of life.

A. Jupiter has four large moons that were discovered by Galileo in 1610: Io, Europa, Ganymede, and Callisto.

B. Io is the most volcanic place in the solar system; with constantly erupting volcanoes of lava and liquid sulfur that turn the surface yellow.

C. Europa is made of rock, but also has a large saltwater ocean beneath a frozen icy crust. This icy crust is very cracked, permitting water to come to the surface. It is the most likely place in the solar system, other than Earth, where life could exist.

D. Ganymede is half rock and half ice. It is the largest planetary moon, and has a crust that is 4 billion years old. Ganymede has a reduced magnetic field that is probably the result of a layer of salt water beneath a thick icy crust. It is less likely to have life than Europa, but is still a possible candidate.

E. Callisto is also half rock and half ice, but smaller and farther from Jupiter than Ganymede or Europa, and therefore has insignificant tidal heating. NASA considers Callisto a good possibility as a base for future outer solar system exploration.

F. One of the most unusual places in the solar system is Titan. It is the largest of Saturn's moons and the second largest of the solar system. Titan, the only moon with a dense atmosphere, 50% thicker than Earth's atmosphere and like ours, is largely nitrogen. The atmosphere contains hydrocarbons like methane and ethane, and the surface has lakes and rivers of liquid hydrocarbons. Titan has changing seasons and climates, clouds, wind, rain, and surface features similar to those of Earth. However, the temperature is very cold (about -180°C), so it is unlikely to have life, but might be a good place to stop and refuel a rocket ship.

V. There are many other smaller objects in the solar system that are not planets or moons—dwarf planets like Pluto, centaurs, trojans, trans-Neptunian objects, comets, and meteoroids. They are mixtures of metal, rock and ice.

A. Pluto used to be a considered a planet, but was demoted in 2006 to the status of a new category: dwarf planet.

B. A planet has now been defined as requiring to orbit only the sun, be large enough to be a sphere, and uniquely occupy the space in its orbit. Pluto does not meet the last condition, as it swings inside the orbit of Neptune.

C. Pluto is part of the Kuiper belt, which, along with the scattered belt objects, contain many planetoids extending out to about 55 astronomical units (AU). One AU is equal to the distance from Earth to the sun.

D. Many tiny icy comets comprise the diffuse Oort cloud, which extends out past 100,000 AU. When these comets get deflected into the inner solar system they lose their ice making long tails that can be spectacular, stretching for many millions of kilometers.

VI. Is there anything else in our solar system that looks like Earth? Is anything alive? Is there any place where we could live?

    A. Earth has had a liquid ocean at the surface, plenty of heat and nutrients, and it still took 3 billion years for multicellular life to evolve here.

    B. It is unlikely that anything other than single-celled life could exist anywhere else in our solar system, although we have a sun that supports complex life on Earth.

    C. Many places have water and rock to support life, and some might even support human colonies.

        1. We will need a source of energy. Nuclear power could be suitable, but is in limited supply.

        2. Tidal heating would be adequate on planets like Io, but Io lacks sufficient sunlight.

        3. Mars would be suitable if there is water that could be found underground.

    D. Living farther from the sun would require giant space-based solar reflectors to concentrate the weak sunlight.

Recommended Reading:

Morrison and Owen, *The Planetary System.*

Robinson, *Blue Mars.*

———, *Green Mars*

———, *Red Mars.*

Questions to Consider:

1. Do moonquakes within Earth's moon and volcanism on Jupiter's moon Io both have the same origin? (Clue: Io is the closest of Jupiter's moons.)

2. What do you see as being the most important planetary characteristic required for the existence of life?

Lecture Forty-Seven—Transcript
The Solar System—Earth's Neighborhood

Welcome. I think that, by this point, you probably have a pretty good understanding of the remarkable and complex factors that have gone into making our planet. And, from the last lecture, the incredible component of life, the whole biosphere, including humans, creatures that cannot only explore their world, underground, and outer space, but write and teach about it to others, for them to share. In the next two lectures, I want to step back and review what it means to be planet Earth. I want to review the many remarkable geologic processes that have not only created our planet, but continue to work together as if in some complex dance, or a multi-dimensional loom, weaving together threads of all different kinds, coming from all different directions to create the continuing fabric of Earth's evolution.

I want to do it in the context of answering a simple question, however: Are there any more out there? I mean, are there any more planets like Earth? As we travel out into space, are we going to find any other planets that we could live on? Are we going to find anyone else there composing music and poetry? These are very different questions, but both very interesting.

In this lecture, I'm going to start close to home in our own neighborhood, in our own solar system. After all, life exists here on Earth, so we know that our sun is clearly a suitable star to foster the evolution of life. In the next lecture, I'm going to broaden out my search to include the whole galaxy, to look at what conditions would be required to allow for the evolution of some other creature that would also be, at this very moment, trying to figure out if we might exist.

Now that we know what it takes to make an Earth, and to keep its geology active enough, and its climate warm and stable enough, to support life for billions of years, what are the chances of their being life—others like us—on other planets like ours? Let's start in our own neighborhood. I'm going to start with Mercury and work my way outward. And you'll see that each of the worlds in our solar system is very different from the others, but many share very similar aspects of geology, things like sand dunes, ice caps, volcanoes, rivers, hurricanes—things that are found on Earth are also found on other planets in our own solar system. Remember that what we're looking for, if we want to find complex life there, or if we want to go there someday, is the combination of just the right kinds of rock, and water, and air. It doesn't sound like too much to ask. Well, let's take a look.

The four inner planets in our solar system—Mercury, Venus, Earth, and Mars—are called the "terrestrial planets," and they are all very similar to each other. Interestingly, they are all very dissimilar from the other four planets in our solar system—the gas giants Jupiter, Saturn, Uranus, and Neptune. But the gas giants have moons, and those moons turn out to be like smaller versions of terrestrial planets, sort of. And then there's a whole lot of other stuff, a whole lot of smaller objects in our solar system as well.

Well, let's start with Mercury. Mercury is mostly iron. There's an iron core, fairly large, a rocky mantle, and a crust, like all the terrestrial planets. Because it formed near the sun, where conditions were hot, it ended up being able to form mostly from iron and not hold on to the lighter materials, the gases, the ices, even not very much rock. Remember when I talked about how the whole solar system formed early on in the course. Mercury, interestingly, has a very strong magnetic field for its size. That means it probably has a molten outer core, just like Earth. However, Mercury is small, very small; its radius, 2,440 kilometers, is less than the size of Earth's core by quite a bit, though it actually turns out to be a little bit larger than our moon.

So, if Mercury is small, it has a weak gravity, which means it has no atmosphere and, therefore, a lot of craters. And that's what you see all over the surface of Mercury, enormous numbers of craters pock marking the whole planet. Why? Because an atmosphere burns up meteoroids as they pass through it. That's what a meteor streak is in our atmosphere. It's a small piece of rock, or metal, or even ice of comet, burning up as it streaks very quickly through our atmosphere. It's interesting: On Mercury, some of these impact basins are huge, so clearly it got hit by some very large objects.

The largest is the Caloris basin; it's about 1,300 kilometers across. Interestingly, on the other side of Mercury is a large area of rifted, fractured terrain. We think what happened was that the impact was so large that it caused seismic waves to spread all the way around the planet and converge at the back end, and essentially, it blew the back end of the planet off. This would have happened during that late heavy bombardment phase, 4 billion years ago or older. That means that the planet Mercury is essentially dead. Its surface has not changed in more than 4 billion years, other than accumulating more and more of these craters. In addition, its interior is probably unchanged as well, so Mercury is not going to be the sort of place that we would want to go.

Now, it's an interesting question: Why is Mercury so geologically inactive when compared to other planets, like Earth? It turns out that small objects cool off a lot more quickly. The heat has a shorter distance to go to reach the surface and escape. Another way to think of it is that it has a larger surface area to volume ratio. It's interesting: If you take two objects that have the exact same shape, the smaller one will actually have a much larger ratio of surface area to volume. If you don't believe it, check it. You can do it for homework. Take two cubes, one meter on a side and two meters on a side. Calculate their areas and their volumes and take the ratios. You'll see that they're not the same.

Size matters. It controls the rate that the planets cool off and other things as well. It turns out, in biology, that the principle of ratio of area to volume is incredibly important. It's why elephants have large ears. Elephants are large; they have fairly small surface area to volume ratios, so it's hard for them to cool off. The ears increase their surface area, allow the heat to leave their bodies more quickly. They're sort of like radiators. It's all the same principle.

Mercury, being so close to the sun, where it's so hot, also means it has no water. And if it has no water, it has no life. You would never want to be there. The temperature of Mercury, as it moves, changes incredibly. It's −175°C on one side and 425°C on the other.

Interestingly, Mercury has a very unusual orbital pattern. It's closest to the sun, so it orbits the fastest, revolves the most quickly of any of the planets. One year going around the sun only lasts 88 Earth days, but one day on Mercury actually lasts 59 Earth days, and the rotations and revolutions are locked in a particular integer ratio of three to two. In other words, Mercury rotates three times for every two revolutions it makes around the sun. It turns out that as we go through the rest of the solar system, that sort of pattern of an integer variation between the revolutions and the rotations is very common. It's called a "tidal resonance." Earth's moon has that. The same side of the moon always faces the earth, so it has a one to one ratio. It rotates one time for every one revolution.

Let's move out. Let's go to Venus. Venus is very close to Earth in size. Its radius is a little more than 6000 kilometers, 95% that of Earth, which means that its volume turns out to be about 86% of Earth, so the force of gravity is about 90% that of Earth's. You'd weigh 10% less. It'd be a great way to instantly lose some weight. So, maybe Venus is a better option as a place to go, maybe to live or to find life. The planet Venus is still tectonically active. It still has volcanoes. Not too many, though. These volcanoes probably resurface the planet on average every half a billion years. It has large gravity, so it has an atmosphere, and because there's volcanism, we know that there's carbon dioxide in the atmosphere, so it's warm. There's a greenhouse effect there.

But here's the catch: It's too warm. The atmosphere in Venus is 95% carbon dioxide, and the surface temperature is 460°C. It's still too close to the sun. It's way too hot, and there's no water. There's no liquid water, and there's no ocean, and therefore, there's no life, so there's no way to pull the carbon dioxide out of the atmosphere. The atmosphere, as a result, is huge. The pressure of the atmosphere at the surface is 90 bars. That's 90 times the pressure of Earth's atmosphere. It's equivalent to being under almost a kilometer of water. Venus has a huge runaway greenhouse. In addition, within its atmosphere, it's got droplets of sulfuric acid. These make thick clouds; we can't even see the surface. We need to use radar mapping to figure out what Venus is made out of.

Well, volcanism also means there's mantle convection, but there are no signs of plate tectonics. Why? Think about it. If you have no water, you have no asthenosphere, and on Earth, it was the asthenosphere that was vital in plate tectonics. It provided the lubrication in the mantle that allowed the stiffer plates to move over it—no water, no asthenosphere, no plate tectonics.

We know that Venus has an iron core and probably even a liquid iron outer core, but Venus doesn't have any magnetic field. Probably because it hardly rotates. Very unusual—Venus actually counter-rotates once in 243 Earth days. Perhaps what you need to get a magnetic dynamo to generate a magnetic field is rotation, in order to have a significant Coriolis effect, in order to cause gyres of fluid in the liquid outer core, in order to move the material in ways that would generate a magnetic field. So, no water, 90 bars of pressure, 95% carbon dioxide, 460°C, sulfuric acid in the atmosphere—there is no life, and I guarantee we will never want to live there. Never has a planet been so poorly named. How does the song go, "You're my Venus, you're my fire?" Well, I don't have any desire, and neither do NASA or any of the programs doing exploration in our solar system.

Earth has a moon, a fairly large one, that formed at the same time that Earth formed. It has a similar composition; it has a mantle and a core, it has a crust—what we call a "regolith." It's actually comprised of shattered rock from impacts on the surface. But Earth's moon is quite small, as far as planets or planetary volumes go, and therefore, it can't hold on to an atmosphere. Gravity on the moon's surface is one-sixth that of Earth's. It's just not large enough to hold onto gases.

Like Mercury, the surface of the moon has largely been unchanged for billions of years. It shows the remains of thousands of impacts that have occurred over billions of years. Now, when you look at the moon, you see both white and dark areas. Those white areas are the highlands; they occupy most of the planet's surface, and those rocks date back to about 4 billion years. The dark areas are called "maria." They formed a little bit more recently, 3.5–3.0 billion years ago.

Now, it's interesting, "maria" is the Latin word for oceans, and boy, was that wishful thinking! The maria, in reality, are flat, flood basalts during a period of tremendous volcanic activity that ended 3 billion years ago. There's been no water found there. It's bone dry. And we should know, because the moon is the only planetary body that we've visited. We've brought rocks back, and we know what the composition of the surface is. So, no water, no life—and unfortunately, if we don't find water on the moon, that means probably no human colonization either, because frankly, it would be really hard to have to bring water to some other place to live. We' really rely on it being there if we would ever live there.

Now, we move out to Mars. This is a slightly different matter. Mars is still small; it's about the size of Earth's core. The gravity on the surface of Mars is only about one-third of Earth's. It's got, therefore, a very thin atmosphere, and it's cold. But it's not that cold. Temperatures on the surface of Mars range from minus $-87°C$ to about $-5°C$, just less than the point of freezing. So, it's sort of like Antarctica, almost livable.

Mars has ice caps. Are they water? No, they're carbon dioxide, because the atmosphere, like Venus, is 95% carbon dioxide. Well, plants would be happy, at least, and Mars does have quite a few Earth-like environments. In addition to the ice caps, it has deserts, sand dunes, dust storms.

It's still a tiny bit tectonically active. There are some bits of evidence that there has been some volcanic activity in the not-too-distant geologic past, though it doesn't seem to be occurring currently. It never had plate tectonics, but it does have signs of crustal deformation that occurred in the past. For example, there's a massive rift called Valles Marineris, a huge crack in the planet that would actually span from New York to Los Angeles—a huge valley. And it shows evidence of massive volcanism 3.5–3.0 billion years ago. In fact, the volcano Mount Olympus is the largest volcano in the solar system, and it would actually reach from New York to Cleveland. It forms in a region of lots of volcanoes called the Tharsis Bulge. Probably what happened was there was a large hot spot in the mantle of Mars. But because Mars doesn't have plates and the surface wasn't moving, the hot spot just kept pumping magma into the same spot on the surface, creating a large number of volcanoes that create a significant lopsided bulge to the planet.

However, to have that much volcanism, you probably had on Mars, at one time, a thick atmosphere and probably even an ocean hundreds of meters thick. What makes people think that is because everywhere on the surface of Mars you see evidence of water—dry lake beds, stream beds, stream erosions. There's no liquid water at the surface now, but there's some evidence that it actually might still be leaking out of the sides of cliff walls. We see evidence of water molecules still leaking out into space. So, is it there, frozen underground, perhaps? This is a very important topic that the NASA program is trying to investigate.

In 2004, in January, one of the most successful missions of NASA ever involved putting two rovers, Spirit and Opportunity, on the surface of Mars. They were designed to operate for three months and actually ended up working for years. They were actually co-led by Ray Arvedsom in my department at the Washington University in St. Louis. In fact, much of my whole department packed up and went to Jet Propulsion Labs in California in order to help steer these things around the surface of Mars. These things went on all sorts of adventures. They got stuck and unstuck in sand dunes, they had to last through dust storms, drove for kilometers on the surface, in and out of craters, and what they found everywhere was evidence of water. Sedimentary rocks that would have formed in an ocean setting, various forms of the mineral hematite that would form in water, chemical weathering aided by water. Was there life? Very possibly, if there was liquid water there. Now, could we live there? Again, the key is that we would need water. We would need to discover water somewhere beneath the surface of Mars.

As you move out from Mars, next you get to essentially a rubble heap that we call the asteroid belt, countless small asteroids between the orbits of Mars and Jupiter. There are more than a million of these that are a kilometer or larger in size. In fact, we've named 160,000 of these asteroids. The largest, Ceres, is considered a dwarf planet, but it's still tiny. In fact, if you took the whole mass of all the asteroids and put them all together, they still would only make up 4% of the mass of the moon. It's not a lot of material.

We used to think that the asteroids formed from a smashed planet, but now we think that probably these are remnants of the early solar system. These are planetesimals that were never allowed to become a planet due to the strong gravitational force of Jupiter. Now, these asteroids are closely monitored, because some of these occasionally swing into the orbit of Earth and have caused fairly significant changes in the history of life on our planet, as with what happened 65 million years ago at the start of the Cenozoic era. However, these small objects have no water, no life, and not much interest for us in terms of a place that we would live. Interestingly, when you go out beyond them, you get to the four gas giants, and also those are not a place where we would expect to find life or be able to support life.

The giant planets Jupiter, Saturn, Uranus, and Neptune are sort of like planets with thyroid cases. I mean, they grew like other planetesimals at the start of the solar system, but they formed just beyond the condensation point for ices. In other words, they could

©2008 The Teaching Company.

begin to accumulate materials like water, ammonia, and methane (which we call ices in the context of the solar system, because they're usually frozen). So, they formed with rock and metal like the terrestrial planets, but also with water, ammonia, and methane, and they got really large really fast as the solar system formed. They became large enough to hold on to the abundant light gases of hydrogen and helium. In fact, if Jupiter were about 50 times larger than it is, it would have been a star in its own right.

It turns out that Jupiter and Saturn are quite similar to each other, and Uranus and Neptune are similar to each other, though there are some significant differences between them. Jupiter and Saturn are mostly made of hydrogen and helium. You don't have any surfaces. The gas transitions directly to a liquid at high pressures. In fact, the name "gas giant" is really a misnomer. They are mostly made of liquid hydrogen and helium. They have weather. They have methane and ammonia in the atmosphere that form these colorful bands that can change a little bit as they move. Those separate bands that you see are due to a very strong Coriolis effect, just like the Hadley cells on Earth. However, Jupiter and Saturn rotate really quickly. Jupiter goes through one day in only 10 hours, and because it's so large, the velocity at any point on its equator is about 45,000 kilometers an hour, whipping around the axis of Jupiter. That causes a very large Coriolis effect and very many different bands of atmospheric circulation.

You get out to Uranus and Neptune, and their composition is mostly that of ices in the mantle, water, methane, and ammonia, with smaller hydrogen and helium atmospheres. Now, each of these has very interesting individual characteristics. Jupiter is the largest, about 11 times larger than Earth, but its volume is, therefore, about a thousand times larger. There's a large red spot that is actually a storm. It's like a hurricane, only it has lasted a tremendously long period of time. As long as we've observed Jupiter, it's still had that red spot.

Now, Saturn is interesting. It has, actually, the lowest density of any of the planets. Its density is less than a gram per cubic centimeter, which means that if you could possibly put it into a giant bathtub, it would float. Saturn is best known for its rings, which are made of rock and ice, and though they look quite substantial, are incredibly thin. They're only tens of meters thick. To try to give you a scale, imagine that you had a round piece of paper to represent those rings.

That piece of paper would have to be 2 miles across in order to have proportionately the same thickness as Saturn's rings. Uranus also has noticeable rings. In fact, all the gas planets do. The most unusual thing about Uranus is that it's actually tilted on its side a little bit more than 90°. We still are not sure how a planet of that size could be tilted, but it's thought that perhaps a very large impact with another planet early on in its history actually knocked the planet on its side.

The farthest of the planets is Neptune, which is 30 astronomical units away form the sun. (One astronomical unit, by the way, is equivalent to the distance between the earth and the sun, and it's 30 times that distance.) Because it's so far out, it takes a long time, a full 165 Earth years, to make one revolution around the sun. Interestingly, Neptune has a tilted axis that's about 29°, not too different from the Earth. And as a result, it has seasons. It actually has changes in its atmosphere probably due to the change in the direction of that tilt of the axis with respect to the sun. For instance, when Voyager II came by in 1989, there was a very large storm spot seen in its atmosphere that was entirely gone by 1994.

Well, in our search for other Earths, there's just not much here. Any rocky surfaces on the gas giants are thousands of kilometers beneath thick mantles and atmospheres of ices and gases at unbelievable pressures. But they have moons, and lots of them, with some very interesting geology. Jupiter has four large ones and many smaller. They were discovered by Galileo in 1610 and actually brought about an end to the whole geocentric model. They are Io, Europa, Ganymede, and Callisto.

Io is interesting: It's the most volcanic place in the whole solar system. It's constantly erupting. You have tidal heating from Jupiter stretching and compressing the planet, creating molten rock and sulfur. We have huge volcanoes of sulfur making the moon turn yellow, in fact. It's probably the worst smelling place in the solar system, except that you don't have an atmosphere because it's too small. You wouldn't smell anything.

The next moon out, though, is the most interesting from our perspective. Europa, like other large moons, is made of iron rock, but also it has a large saltwater ocean beneath a thin, frozen, icy crust. The water there is probably 100 kilometers thick, or so, with an ice layer that may be up to 10 or 30 kilometers. It could be much less.

What's interesting is that the icy crust is very cracked, and sometimes water actually comes up to the surface to fill the cracks. Again, it's the heat that drives this process as a result of the tidal forces from Jupiter. And because we've got liquid water, sometimes at the surface, it's the most likely place in the solar system, other than the Earth, that life could exist.

The other moons are interesting also. Ganymede, which is half rock and half ice—actually, it's the largest planetary moon—is made of rock and mantle but also has an icy crust. The crust is old, about 4 billion years. We see signs of old impact cratering. But it has an induced magnetic field that probably is the result of a layer of salt water beneath the icy crust. It's deeper down than Europa's, so it's less likely than Europa to have life, but it still is a possible candidate.

Callisto, the next moon out, also is about half rock and half ice, but it's smaller, farther from Jupiter than Ganymede. It's about the size of Mercury, and it therefore has fairly insignificant tidal heating, and as a result, the moon just never separated. It never differentiated very well. It may also have a small, deep, liquid ocean, but it's less likely than either Europa or Ganymede. Actually, NASA considers it a fairly good place for a human base for future exploration of the outer solar system.

One of the most unusual places in the solar system is Titan. It's the largest moon of Saturn, the second largest in the solar system. It's the only moon with a dense atmosphere, one and a half atmospheres. It's 50% thicker than Earth's atmosphere, and it's largely nitrogen, just like ours. It has hydrocarbons like methane and ethane. It actually has lakes and rivers. However, it's really cold; it's −180°C, so these lakes and rivers are not water. They're liquid hydrocarbons. However, it has climates. It has clouds; it has wind and rain that create surface features similar to that of Earth—sand dunes, shorelines. It even has changing seasons. It's sort of like what Earth would have looked like, an early Earth, but under much colder conditions, so cold that it's unlikely to have life, but it would be a great place to stop and refuel your rocket ship.

There are many other moons and many other smaller objects in the solar system that are not planets—dwarf planets like Pluto, centaurs, trojans, trans-Neptunian objects, comets and meteoroids. They're all mixtures of metal, rock, and ice, though the farther you go out, the less metal and rock and the more ice that you have.

Pluto, of course, used to be a planet, but it was demoted in 2006 to a dwarf planet status, because it no longer meets the criteria for a planet. A planet has to orbit the sun and only the sun, not be a moon of something else. It has to be large enough to be a sphere, and it has to clean out the region around its orbit. Pluto meets all those except for the last. It swings inside the orbit of Neptune, and so is no longer considered a planet. Besides, Pluto's small. It's actually smaller than many of the moons around other planets. Pluto's part of a Kuiper belt that extends far out into the solar system—lots of objects, more than 70,000 objects probably larger than a kilometer in size, going out from Neptune to 55 or so astronomical units away.

If you go even farther out, you get to the Oort Cloud of comets, that goes out to 100,000 astronomical units—probably millions of small bits of ice that sometimes get deflected into the inner solar system, where we see them as bright streaks across the sky, the comets. But their total volume is very small. There are lots of them, but they are really small ice balls, snowballs. In fact, when Halley's Comet came through in 1986, we were able to see that even though its tail stretched for millions of kilometers, the comet itself was only about 15 kilometers across.

So, what are our prospects, here in our own local solar system? Well, is there anything else that looks like Earth? Anything alive? Anyplace that we could live? There are several places with liquid water, but not right at the surface. They are cold, harsh environments. There do seem to be some moons with liquid water in an ocean beneath the surface, and life could exist there. Again, anywhere that we find liquid water on the surface of Earth, we find life, but it would be a hard life.

Earth has had a liquid ocean at the surface, lots of heat and nutrients, and it still took 3 billion years for multicellular life to evolve here. I don't expect anything other than single-celled life could exist anywhere else in our solar system, even though we know we have a sun that could support complex life like what exists on Earth. Nothing that's going to come visit us one day—no Martians and certainly no Venutians.

Could we live there? Maybe. Lots of places have water and rock to build and live by. We'd need a source of energy. Nuclear power works, but uranium is in limited supply. Tidal heating works in places like Io to keep the geothermal energy going, but Io is not a very friendly place to be; you really want to have sunlight. Mars would work, if there's water underground. There are even lots of other things that would remind us of home, like sand dunes, sedimentary rock, and even ice caps. To go out in the solar system, far from the sun, you would need some space-based solar reflectors to concentrate the weak sunlight, some way to capture sunlight. But you could do it. There's lots of rock and ice, water ice, and Titan even has clouds, and weather, and rivers, and lakes, and a whole lot of petroleum, as well.

Well, in the next lecture, the last lecture, I'm going to extend my search outward. What are the chances of finding other Earth-like planets in the rest of the galaxy? In the process, we'll review what the properties are that have been vital in making Earth the remarkable place that it is, geologically and biologically.

Lecture Forty-Eight
The Lonely Planet—Fermi's Paradox

Scope:

What are the chances that there are other civilizations in our galaxy that, at this moment, are wondering how many other civilizations there are in their galaxy? Many scientists used to think that the number was high. However, recent geophysical and astrophysical research suggests that the conditions required to have liquid water continuously available at a planet's surface for billions of years, which was needed for us to evolve, may be quite rare. Earth needed to be the right size, with the right mix of compositions, the right kind of moon, the right-shaped orbit, and the right distance from the sun. The solar system needs to have planets like Jupiter and Saturn, with nearly circular orbits, to keep asteroids from continuously bombarding Earth yet not destabilize the orbits of other planets and the asteroids. Our sun needs to be the right size star, in the right location, and in the right kind of galaxy. Put all of these requirements together, and it suggests that there may be only a handful of planets in the Milky Way galaxy like Earth, if any others at all. Life is probably common, starting up wherever there is liquid water. But the stable conditions required for the evolution of something that becomes aware of its own existence may actually be unique to Earth.

Outline

I. In 1950, the famous physicist Enrico Fermi was discussing recent experiments about the origin of life with his colleagues. It seemed that life might start on any planet that had liquid water and an atmosphere. Fermi asked a simple question that has puzzled scientists ever since: Where is everybody? This became known as Fermi's paradox.

   A. Given the size and age of the universe, there should be other civilizations in existence, but no one has contacted us from any other planet.

   B. We know there are many planets. In fact, in 2007, the first Earthlike planet, Gliese 581c, was discovered only 5 light years away from us. But does it have intelligent life?

C. How many of these planets we are discovering have life? In how many cases is that life complex, multi-cellular or intelligent? Is anybody out there? If so, where is everybody?

II. There have been many attempts to answer Fermi's paradox.

 A. Maybe other civilizations have tried to contact us but we don't recognize the signs.

 B. Maybe we aren't looking in the right places, or at the right things? Maybe the creatures are too alien to be able to communicate with us? Maybe they all eventually choose to be nontechnical?

 C. Maybe civilizations don't last long? Maybe it is the nature of intelligent life to destroy itself? Or to destroy others?

 D. Maybe they are intentionally avoiding contacting us? This is called the "zoo hypothesis."

 E. Perhaps they just aren't there. Earth is a remarkable planet, and maybe there are few like it.

  1. Earth's processes are delicately and sensitively balanced with a solar heat engine shaping the surface and driving climate change, water flow, weathering and erosion within narrow fluctuations in temperature.

  2. Maybe the requirements to make a planet like Earth are remarkably stringent.

 F. There was a time when scientists calculated that there might be numerous civilizations in the galaxy.

  1. Frank Drake, a famous astrophysicist, in 1960 constructed a formula designed as a thought problem to try to address Fermi's paradox. His Drake equation took the form of a product of probabilities that would give the number N of inhabited planets.

  2. Carl Sagan, the famous astronomer and author, said that because there were billions and billions of stars in our galaxy (it's now thought to be between 200 and 500 billion), and there are more than 100 billion galaxies in the universe, there must be trillions of civilizations.

III. It is remarkable that we even have planets, because the very existence of stars and planets requires extremely narrow bounds on the fundamental laws of the universe.

A. If the relative strengths of the four fundamental forces of the universe were only slightly different, planets and people and all of matter could not exist. This is known as the Anthropic Principle, or Goldilocks Enigma.

B. A wide variety of solutions have been put forward in order to explain the Goldilocks Enigma.

    1. According to the absurd universe, it just happens to turn out this way (by random chance).

    2. According to the unique universe, there is a deep underlying principle of physics that requires the universe to have worked out this way; some theory of everything that we just haven't found yet.

    3. One proposal is that we are living in a fake universe; a kind of virtual reality.

    4. Some have proposed a universe designed by an intelligent creator specifically to support complexity and the emergence of intelligence. However, this does not address the troubling question of who created the creator, and we have to go through the whole analysis again, replacing "universe" with "god."

    5. A popular solution is that there are multiple universes existing in parallel. Many physicists favor this because it is one outcome of string theory. Because this model is not currently testable, however, it doesn't really fall within the realm of science.

IV. Recent work suggests that even in a universe favorable to the formation of planets, like our universe, the conditions required to support complex life on a planet may be incredibly small, and Earth just happened to have the right conditions. This presents a whole new Goldilocks Enigma.

    A. Earth is in the right location in our galaxy.

        1. Most stars in a galaxy are close to the center, but nearby star passes would throw trillions of comets into any planets causing very high rates of bombardment. In addition the intense radiation and frequent explosions of large stars would prevent the existence of life near galactic centers.

        2. At the edges of galaxies, star light suggests that star systems are starved for metals, with very low levels of silicon, iron, magnesium, and all the other requirements for planets and the building blocks of life. This is because stars are small so they last a long time, and there have not been enough supernovae to produce the heavier elements needed to make rock.

3. Outer galaxy stars also show low amounts of radioactive elements, so even if there were planets they would be geologically dead.

4. There is only a narrow zone in spiral galaxies, maybe 5%–10% of the total number of stars, that is sufficiently metal-rich and could support life.

5. Many galaxies, such as elliptical and irregular ones, seem to be metal-poor and therefore likely unable to support life. Only spiral galaxies (like ours) are metal-rich but, even then, only near the centers.

B. Our sun is just the right size.

1. Large stars have very short lifetimes and emit too much ultraviolet radiation. The habitable zone for our sun is about 5% closer than Earth is now, and 15% farther away.

2. For smaller stars the habitable zone is closer to the sun because they emit lower levels of energy. However, planets close to the sun risk danger from solar flares. They also tend to be tidally locked so that one side always faces the sun and burns; the other always faces away and freezes.

C. Jupiter is just the right kind of shepherd.

1. Jupiter prevents Earth from getting bombarded by probably 10,000 times the number of comets and meteoroids that actually do hit it by flinging them out into space.

2. If Jupiter's orbit were slightly more elliptical or if it were much larger, it would have the opposite effect and gravitationally destabilize Earth's orbit and the asteroid belt.

3. All of the Jupiter-sized planets that have been observed so far in other solar systems are bad Jupiters with very eccentric orbits.

V. Earth seems to be the right kind of planet to support life in terms of its location in the galaxy, its sun, and its own geology.

   A. Earth is large enough to hold onto a thin atmosphere, but not too large to be smothered with hydrogen and helium like the gas giants.

   B. Earth has the right balance of rock and metals.

   C. Earth has a nearly circular orbit that keeps it at just the right distance from the sun to maintain liquid water.

   D. Earth's large moon is at just the right distance from Earth to behave like a gyroscope and minimize the changes in the tilt of Earth's axis. This keeps the Milankovitch cycles small enough to maintain a relatively stable climate.

   1. The formation of the moon is important for giving Earth a fast rotation rate which keeps day and night temperature swings from being too great.

   2. The moon also provides Earth with its steadily tilted axis, which gives us seasons. Seasonal climate variations have been an important stimulus for natural selection and the evolution of life on the planet.

   3. When the protomoon hit Earth, Earth absorbed most of the protomoon's iron core, making Earth's core larger, and giving it a strong magnetic dynamo and a large magnetic field. That magnetic field produced the magnetosphere which protects Earth from solar wind and solar particles that are constantly shot out at us from the surface of the sun.

   4. Earth has sufficient carbon to aid in the development of life. It has both oceans and land mass enough to balance and regulate that carbon cycle.

   5. Earth has enough radiogenic isotopes to keep it warm and geologically alive with mantle convection, plate tectonics, air, land, water and the many ecological niches and microclimates that have promoted tremendous biodiversity.

   E. In short, Earth seems to be just the right size, just the right distance from just the right kind of star, and just the right distance from the center of just the right kind of galaxy. Intelligent life on Earth may have occurred in spite of overwhelmingly strong odds against it.

VI. In light of the many challenges to providing stable conditions for life on the surface of a planet, maybe we are alone. If so, what are the implications of this?

  A. For example, if there is just one creature in the universe that resembles a tiger, do we have an added responsibility to make sure it does not go extinct?

  B. It seems that we humans are maturing as a culture and are beginning to realize both our power as a geologic force and the responsibility that comes with that power.

  1. Einstein once wrote, "A human being is a part of the whole called by us the universe, a part limited in time and space. ... Our task must be to free ourselves ... by widening our circle of understanding and compassion to embrace all living creatures and the whole of nature in its beauty."

  2. Our cultural evolution has been one of continually widening the circle of compassion, from individuals, to families, to extended families, to a clan or pack, to a community or town, a city, a state, a country, and a continent; perhaps eventually the planet, solar system, galaxy, universe, and maybe even multiverse.

  C. The philosopher Alfred North Whitehead once equated complexity with beauty and viewed maximizing this beauty as equivalent to reaching God. Perhaps even the Internet can be seen as part of this process; an increase in complexity and therefore a kind of beauty. Perhaps all that we are doing on Earth represents the next stage along the long process that started with the Big Bang and moved from energy, to matter, to life, to mind, and next, maybe to spirit.

  D. If any advice can be given from a doctor of geology, it is to get out and enjoy the world.

  1. For example, you will see a beautiful and majestic mountain now also as a battleground between the forces of tectonics that made it, and the forces of erosion that are tearing it down.

  2. A shoreline will not just be a soothing and relaxing walk in the sand, but a vibrant, pounding struggle as the waves continuously try to make crooked coastlines straight again. You will also see it as a factory for the sedimentary rocks that you will find elsewhere in places like the Grand Canyon.

3. Marcel Proust said that the real voyage of discovery consists not in seeing new landscapes, but in having new eyes. Even better: You will have new eyes with which to see new landscapes.

Recommended Reading:

Davies, *The Goldilocks Enigma.*

Ward and Brownlee, *Rare Earth.*

Webb, *If the Universe Is Teeming with Aliens ... Where Is Everybody?*

Questions to Consider:

1. Imagine a planet slightly larger and hotter than ours. What kind of life do you think would evolve to be the "intelligent" life form on it?

2. Aliens are often portrayed in science fiction as being fairly similar to humans. Given the success of vertebrates on Earth, do you think that the dominant life forms on planets will always evolve to be quadruped vertebrates?

Lecture Forty-Eight—Transcript
The Lonely Planet—Fermi's Paradox

Welcome. One day, back in 1950, the famous physicist Enrico Fermi was having lunch with some colleagues, and they were talking about recent experiments about the origin of life. Remember those discoveries that life would begin naturally in the presence of water, carbon dioxide, ammonia, and some of the other basic materials that were found in Earth's early atmosphere? Well, it seemed that life might start up on any planet that had liquid water and an atmosphere. So maybe Earth was just an ordinary planet, but Fermi paused and then asked a simple question that has puzzled scientists for more than a half century. "Well, where is everybody? Given the size and the age of the universe, there should be all these other civilizations, then, right? Why haven't they contacted us?"

That became known as Fermi's paradox. We search the galaxy with visible light and radio waves, and we just don't see any signs of life. Surely, if there had been all these civilizations around for billions of years, at least one of them would have sent a probe here. I mean, you can get most anywhere in the galaxy within just a few tens of millions of years, and Earth would be an obvious candidate for life. They could tell that from the surface temperature of our planet.

We know there are lots of planets around other stars. Actually, using several very clever techniques, astronomers have now discovered several hundreds of them. In fact, in 2007, the first Earth-like planet, Gliese 581c, was discovered orbiting a red dwarf star, Gliese 581, that's only about 20 light years away from us. But does this planet have life? How many planets out in the solar system actually have life on them, and in how many cases is that life complex, multicellular, or intelligent? Is anybody out there, and if they are, where is everybody?

These are some of the most important questions that face science, because it's an integral part of being human to have that drive and desire to discover who we are, and how we got here, and if we're special. Are we unique in the galaxy, or are we just one of many different organisms that has solved the riddle of consciousness, that's made that long crawl through the mud? Either way, it would totally change the way we think, right? Arthur Clark once said, "Sometimes, I think we're alone in the universe, and sometimes I think we're not.

In either case, the idea is quite staggering." Think about it. Are there lots of other planets like us? Are we one of just a whole, huge community? Or are we it?

There have been a lot of different attempts to answer Fermi's paradox. For instance, maybe these civilizations have tried to contact us, but we just don't recognize the signs. Arthur C. Clarke once said that any sufficiently advanced technology would be indistinguishable from magic. Maybe it's harder to get a message across space than we think, or maybe we aren't looking in the right places or at the right things. Maybe these creatures would just be too alien to even be able to communicate with us. Maybe they all eventually choose to be nontechnical and return back to a life devoid of the industry and complexities that now seem to drive our civilization. Maybe civilizations don't last that long. Maybe it's the nature of intelligent life to destroy itself after some point in time. Or, even worse, maybe to destroy others. Maybe they're intentionally avoiding contacting us; this is sometimes called the "zoo hypothesis."

Well, let me pose another solution to Fermi's paradox. Maybe they just aren't there. During the previous 47 lectures of this course, I've shown you that Earth is a remarkable planet. It's an incredible place. We've got this high energy internal heat engine that drives heat flow that drives mantle convection that moves the plates around as part of plate tectonics. These are so delicately and sensitively balanced with a whole solar heat engine that shapes the surface, and drives climate change, and water flow, and weathering, and erosion, all within such narrow fluctuations in temperature.

We've had liquid water continuously at or near Earth's surface for 4 billion years. That's one-third of the age of the universe. And you know, maybe this just doesn't happen very often. Maybe the requirements to make a planet like Earth are remarkably stringent. Now, Gliese 581c was the first planet out of hundreds so far discovered outside our solar system to even have a surface temperature close to ours. But as I'm going to show you here, there are many factors that all have to line up for us to expect a welcoming party when we eventually go and visit there.

Now, this hasn't always been the conventional view. There was a time when scientists calculated that there might be enormous numbers of civilizations out there in the galaxy. For instance, Frank

Drake, a very famous astrophysicist back in 1960, constructed a formula that's now called the "Drake equation," that he designed as a thought problem to try to address Fermi's paradox. This equation he constructed calculated how many planets there might be within our galaxy that would have civilizations—advanced life forms—that would be attempting to contact us at this very moment.

Now, Drake was one of the founders of SETI, the Search for Extraterrestrial Intelligence, and he encouraged people to look for possible communications from outer space. The Drake Equation that he came up with took the form of a product of probabilities that, when you multiplied them all out, would give you some number, N, of inhabited planets with advanced civilizations, wanting and attempting to contact us. Basic probability works that way. If you want to find the likelihood of any particular event occurring, the odds of that event, you multiply together the separate odds of each particular part. So, for instance, if you want to know what the odds are of rolling boxcars (double sixes) with two dice, well, the odds of rolling a six on one die are 1 out of 6, and the odds of rolling a six on the other die are 1 out of 6. You multiply them together, you get 1 out of 36. That's the odds of rolling double sixes.

So, what Frank Drake did was to take all the different factors involved with having an advanced civilization, and he multiplied them together. Let me show you one form that this equation takes. When N is the number of these civilizations, $N = R_* \cdot f_s \cdot f_p \cdot n_e \cdot f_l \cdot f_i \cdot f_c \cdot L$.

Let me break these down. R_* is the average rate of star formation per year, the number of new stars that are actually forming in our galaxy. f_s is the fraction of stars that would be suitable suns for planetary systems. f_p is the fraction of those suitable suns with planetary systems on them. n_e is the number of planets in the continuously habitable zone. I'll get back to this, what that means, but it's the zone around the star where a planet like Earth could exist, essentially with liquid water on its surface. f_l is the fraction of these planets on which life actually originates. f_i is the fraction of these planets on which life eventually becomes intelligent. f_c is the fraction of intelligent species on these planets that are willing and able to communicate with others. L is the average or mean lifetime in years of a communicative civilization. And again, N is that number of intelligent civilizations within our galaxy that are able and interested in communicating with us.

Well, it's really hard to solve this equation, because we actually don't know the estimates for many of these figures, here. We just have one planet, Earth, to go by. But Carl Sagan, a very famous astronomer and author, said that because there were billions and billions of stars in our galaxy alone (it's now thought to be somewhere between 200 and 400 billion stars in our galaxy), and there are more than 100 billion galaxies in the universe. That meant many, many trillions of civilizations out there.

Well, to begin with, it's remarkable that we even have planets. I talked about this earlier in the course; the very existence of stars and planets requires incredibly narrow bounds on the fundamental laws of the universe. Remember, back in Lecture Six, I said that if the relative strengths of the four fundamental forces of the universe were only slightly different, then planets and people, all of matter, material, couldn't ever exist. For example, if the strong nuclear force were only slightly larger than it is, then all the hydrogen in the universe would have converted to helium right in the early moments of the universe. There'd be no water; there'd be no long-lived stars. Nothing would look like what we have today. Make gravity slightly smaller than it is, and no stars or planets would ever form. The whole universe would just be a diffuse cloud of hydrogen and helium. These sorts of consequences hold for all four of those fundamental forces.

This is known as the Anthropic Principle or, sometimes, the Goldilocks Enigma. It's a paradox of why our universe seems so finely tuned toward having just the right characteristics needed to support planets and people. Now, the Anthropic Principle has taken many forms, and it's often been misinterpreted and misused. But let me give you a couple of solutions, a couple of proposals that people have given for solving this Goldilocks Enigma. Maybe we live in an absurd universe. In other words, it just happens to have turned out this way by random chance. That's possible, but frankly, it's not very satisfying, because the odds against it are just so overwhelming. Maybe we live in a unique universe. In other words, there is some deep underlying principle of physics that just requires the whole universe to work out this way, some theory of everything that will explain why all the various features of the universe must have exactly the values that we see—only we just haven't found it yet.

You know, I'm okay with this. There's so much that we don't know, especially on these big scales of space and time. Maybe there's an overwhelming, underlying life principle, some principle that constrains the universe to evolve towards life, and mind, and spirit, and again, we just haven't found what that is yet. Again, that's not very satisfying. Some people have proposed the fake universe, in other words, we're living in a virtual reality simulation, like in the movie *The Matrix*. The real world has rules that are much simpler and much more obvious, only we don't live there, so we don't see that.

A lot of people have proposed the designed universe, that an intelligent creator designed the universe specifically to support complexity and the emergence of intelligence. In general, there's absolutely no conflict between science and religion or belief in a god. Many scientists believe in God, though in this case, we still have the troubling question of who created the creator, and then we have to go through this whole analysis again on the creation of a god.

Or the multiverse. Maybe there are multiple universes, maybe an infinite number all exist in parallel, and each one of these has slightly different ratios of these fundamental forces. So, in other words, you span the full spectrum of possible combinations of characteristics of universes, and we naturally find ourselves within the one that would support our existence. Interestingly, many physicists actually favor that multiverse interpretation, because it turns out that the existence of many parallel universes is also an outcome of string theory. However, at the moment, this isn't testable, and some people question whether this even falls within the realm of science.

Back to aliens, or the lack of them. Recent work in the areas of planetary biology and astrobiology suggests that even in a universe that is favorable to the formation of planets like our universe, the conditions required to support complex life on a planet may be incredibly small. And Earth just seemed to have those right conditions. It's another Goldilocks Enigma. To begin with, we're in the right location in our galaxy. Most stars in a galaxy are close to its center, and this is a terrible place to live. You've got nearby stars passing by that would regularly throw trillions of comets into any planets. You have enormously high rates of bombardment. In addition, there's been intense level of ultraviolet radiation in the center of the galaxy due to frequent explosions of large stars, supernovae going off all over the place. The planetary surfaces would be highly irradiated as well as bombarded.

Now if you go to the other end of the galaxies, at the edges, you have very few planets. Starlight suggests that star systems out there are starved for metals. There are low levels of silicon, and iron, and magnesium, all the other materials for building planets and people. The building blocks of life—they're just not there, because the stars are so small, they live a long time. You don't have enough supernovae to make heavier elements. You'd also have very low amounts of radioactive elements, and any planets that formed there without the radioactive elements would be geologically dead. Remember that it's the decay of the uranium, potassium, and thorium that keeps our planet simmering on a low boil and that keeps all our planet alive, forming land and generating oceans and atmospheres. You wouldn't have any of that on the edge of the galaxy.

So, it turns out that there's a narrow zone in spiral galaxies that's in the middle. Maybe 5–10% of the total number of stars could even be considered as possible candidates for planets like Earth and the existence of life on them. It turns out that many galaxies, elliptical and irregular galaxies, seem to be entirely metal-poor, likely unable to support life at all. It turns out that only spiral galaxies like ours are significantly metal rich and, even then, only near the centers. In addition, our Sun is just the right size; it's not too small that it would emit too little energy, and it's not too big that it would burn out quickly.

It turns out that large stars have very short lifetimes. Only smaller stars can last for billions of years in their main hydrogen-burning phase, and we need those billions of years for the evolution of life. Large stars also emit far too much ultraviolet radiation. It would be damaging to living tissue. So, it turns out all stars have a very narrow habitable zone. As I said, that's the distance from a star where a planet could exist with its temperature in between the freezing and boiling points of water. For our Sun, it's about 0.95 astronomical units to about 1.15 astronomical units. Remember, an astronomical unit is the distance from the sun to the earth. So, in other words, that habitable zone for our Sun is only about 5% closer than the earth is to about 15% farther. So, it's a very narrow zone that our Earth could have been in in order to have life existing as long as it did, billions of years.

Now, for small stars, like red dwarves, the habitable zone is much closer to the sun, because these stars emit much lower levels of energy. Well, if you're that close to a sun, there's a significant danger of solar flares that could bathe the planet with high levels of very destructive ionized plasma. Also, planets that are close in to a star tend to be very tidally locked, like the way the moon is locked around Earth so that one side always faces Earth. Well, the same thing would likely happen, so you would have one side of the planet constantly facing the sun, and it would be burning. The other side would be constantly facing away from the sun, and it would be freezing.

Well, that's important, because it turns out that small stars are about 90% of all stars, so that may eliminate huge numbers of stars in our galaxies from even being possible candidates. In addition, there are some stars that actually change their energy output very greatly over time, so they would either move in or out of that habitable zone and wouldn't be able to support life continuously for a long amount of time. So, being generous, stars like our sun are probably 5% of all stars.

In addition, Earth, it turns out, has just the right shepherd, Jupiter. Jupiter prevents Earth from getting bombarded by probably 10,000 times the number of comets and meteoroids that actually do hit it, because the gravitational field of Jupiter just flings them right out of the inner solar system and out into space—out of the inner solar system. However, if Jupiter's orbit were slightly more elliptical than it is, it would have just the opposite effect. It would gravitationally destabilize, for instance, the asteroid belt, as well as the orbit of Earth, and the other terrestrial planets. There would be stuff flying all over the place. Well, this turns out to be important, because all the Jupiter-sized planets that have been observed in other solar systems so far are "bad" Jupiters, with fairly eccentric orbits.

So, in all of the other situations that we've observed so far, it would be hard for a planet like Earth to survive the gravitational destabilization of their solar system, and they would be constantly bombarded by smaller objects. Interestingly, if Jupiter was also much larger than it is, it would also be a problem, destabilizing Earth's orbit. So, Earth seems to be the right kind of planet to support life in terms of its location in the galaxy, and its sun, and, it turns out, also its own geology. It's large enough to hold onto a thin atmosphere, but not too large to be smothered with hydrogen and helium like the gas giants.

The large planets attract far too many comets and meteoroids; you get way too many impacts. It's also harder for mountains and continents to form with higher gravity. You'd probably end up with one giant ocean, and you'd lose that land feedback mechanism for regulated carbon dioxide.

Earth has just the right composition. It's got a great balance of rock and metals, and life uses this stuff. It needs all of these materials. In addition, Earth has a nearly circular orbit that keeps it at just the right distances to maintain liquid water. And even so, the ice ages and interglacial periods that we talked about in lectures on climate result from the slight fluctuations in Earth's orbit, those Milankovitch cycles we talked about.

Earth, in addition, has a large moon at just the right distance. This turns out to be important because having the moon orbiting Earth acts like a large gyroscope. It actually minimizes the changes in the tilt of Earth's axis, and that maintains climate stability. So, the Milankovitch cycles are very small compared to other planets. Even so, they cause large climate changes, so it's really important having that small fluctuation and having the presence of that large moon.

The formation of the moon was also probably very important for Earth, because to begin with, it gave Earth its tilted axis, it gave us a fast rotation rate, and it gave us a large core. Now, the fast rotation probably existed from that impact, and keeps day and night temperature swings from being too great. It's just the opposite of having a planet with the same side always facing the sun. Here, because Earth is spinning, the temperatures get quite evened out. If Earth spun more slowly, there would be huge temperature swings day and night. The tilted axis turns out to be important because it gives us seasons, and that may actually have helped evolution by providing climate variations, which have been a huge stimulus for natural selection and the whole evolution of life on the planet. And as far as the core goes, when the protomoon hit Earth, and Earth absorbed most of the protomoon's iron core, that made our core much larger and gave us a large strong magnetic dynamo. That produces a large magnetic field, which produced our magnetosphere, which protects us from the solar wind, from solar particles constantly shot at us from the surface of the sun.

Earth also has enough carbon to aid in the development of life, but not too much so that you end up with a runaway greenhouse effect, like what happened with superheated Venus. Or too little, which would have put us in a permanent ice age. And Earth relies upon both oceans and land to balance and regulate that carbon cycle.

So, Earth has a good amount of radiogenic isotopes as well. As I mentioned before, this keeps Earth warm and geologically alive. It powers mantle convection and plate tectonics. It gives us land, air, and water and creates the many ecological niches and microclimates that have promoted tremendous biodiversity. In short, Earth seems to be just the right size and just the right distance from just the right kind of star at just the right distance from the center of just the right kind of galaxy. It's sobering to contemplate the implications of this line of research.

We can now return to the Drake Equation and try to guess on some of the values. Our star might be about 6. For the fraction of stars that are suitable suns, maybe 1 out of 20. For the fraction of suitable suns with planetary systems, maybe a half; it's at least 10%. For the number of planets around suitable suns in a continuously habitable zone, maybe 1 out of 150—not very high. The fraction of these planets on which life actually originates, I'm guessing probably 1. I think it happens all the time, if you have the right conditions. The fraction of these planets on which life would eventually become intelligent, maybe 1/500. The number of intelligent species on these planets that would be willing and able to communicate with others? I don't know, 50%? That's a total guess. The average or mean lifetime in years of a communicative civilization? I'm being very optimistic, maybe a million years. Multiply them all together, and you would get 1, and that would be us.

Okay, I obviously tinkered with these numbers to have it come out to 1. I don't have any idea what many of these numbers should be, and no one does. It's really only meant as a thought experiment, not a rigorous calculation. The important point is that the number could be very low, even much lower than 1!

In other words, intelligent life on Earth may have occurred in spite of overwhelmingly strong odds against it. Interestingly, when Isaac Asimov wrote a series of futuristic books called the "Foundation Series," he predicted a galaxy with no other intelligent life forms there. This was before the Carl Sagan era, and later, generations of authors criticized him for this, what they called an oversight. In fact, when a trio of younger authors, David Brin, Greg Bear, and Gregory Benford, set out to complete Asimov's Foundation Series, they came up with some scenario where human-made robots had gone out into the galaxy and exterminated all other intelligent life forms to make way for human expansion. That fit their assumptions.

But maybe Asimov was right after all. Maybe we are alone. Maybe there will be no one on Gliese 581c to welcome us when we arrive. What are the implications of this? Will this change the way you view Earth and humanity? If there really is only one creature in the universe that resembles a tiger, do we have a responsibility to make sure we don't drive it to extinction? I don't think we will, because, as I mentioned previously, I think we're growing up as a society, as a civilization. I think we are realizing both our power as a geologic force and the responsibility that comes with such power.

For example, Einstein once wrote, "A human being is a part of the whole called by us the universe, a part limited in time and space. He experiences himself, his thoughts, and feelings as something separate from the rest—a kind of optical illusion of consciousness. This delusion is a kind of prison for us, restricting us to our personal desires and affection of a few persons nearer to us. Our task must be to free ourselves from this prison by widening our circle of understanding and compassion to embrace all living creatures and the whole of nature in its beauty." And you know what? We're doing this. Our cultural evolution has been one of continually widening the circle of compassion, from individuals, to families, to extended families, to a clan or pack, to a community or town, a city, a state, a country, a continent. I mean, look at the European Union forming—eventually, maybe our whole planet, our solar system, our galaxy, the universe—who knows? Maybe even the multiverse.

As I hope these 48 lectures have demonstrated, Earth is a remarkable planet. It's a delicate balance between so many complex and interconnected, different systems—physical, chemical, geological, biological. I have gone through all these different areas. And yet, there's nothing as complex as humans and the world we're creating (perhaps in the galaxy). The philosopher Alfred North Whitehead once equated complexity with beauty and equated maximizing this beauty as reaching God. He said that God is in the future, and we're moving toward God. God is not up there, but up ahead in time. In this way of thinking, even the internet can be seen as part of this process, as an increase in complexity, a kind of beauty. The long process that started with the Big Bang moving from energy, to matter, to life, to mind, and next, maybe to spirit.

Of course, you don't need science or philosophy to know that our planet is beautiful. As Bob Dylan said, "You don't need a weatherman to know which way the wind blows." The only thing that gets painted or drawn more often than the human body is the natural landscape—mountains, oceans, sunsets, fields, forests, rivers, lakes. Think of a place in your past where you have felt most at peace with yourself in the world, a place where you go with your memories to meditate or relax, to feel connected or at one with the world. I bet it's a place in nature—the ocean, the top of a mountain, beside a stream.

And I have to say, if I can give any advice as a doctor of geology, it's to get out and enjoy the world absolutely as much as you can. I mean, Earth has so many different, unique climates in the environment, each one with unique geology and life that has evolved to live in it. You don't have to fly to Alaska or the Himalayas to experience it, although it would be cool. There are parks and places everywhere, and they all have their own different worlds to explore. Remember, you don't need to avoid people; we're a big part of the earth too. I guarantee that you'll see the whole Earth around you in an entirely different way. A mountain will not just be beautiful and majestic, but you'll see it as a battleground between the forces of tectonics that made it and the forces of erosion that are tearing it down.

A shoreline will not be just a soothing and relaxing walk in the sand, but you'll see it as a violent, pounding struggle as the waves continuously try to make crooked coastlines straight again. You'll see it as a factory for all the sedimentary rocks that you see elsewhere in places like the Grand Canyon. I guarantee that you'll never look at a bowl of miso soup the same way. Marcel Proust said that the real voyage of discovery consists not in seeing new landscapes, but in having new eyes. Well, that may be true. You will have new eyes, but get out and see the new landscapes as well.

Timeline

Earth History

13.7 billion years ago .... The universe is created during the Big Bang; three minutes later hydrogen begins forming.

13.6 billion years ago .... The Milky Way begins to form; hydrogen and helium begin to come together to form stars.

4.57 billion years ago .... The Solar System forms; the sun, planets, and countless smaller bodies coalesce into a giant rotating disk-shaped nebula.

4.52 billion years ago .... Our moon forms; Earth is likely entirely molten at this point.

More than 4.4
billion years ago .... Earth's solid crust begins to grow and persist.

More than 4
billion years ago .... The earliest continents form; plate tectonics takes on the same form that is in operation today.

More than 3.5
billion years ago .... Life begins as simple single-celled prokaryotic organisms.

3.3 billion years ago .... The earliest supercontinent, Vaalbara, forms.

2.5 billion years ago .... The permanent presence of oxygen gas in the atmosphere, paving the way for the evolution of animal life.

Less than 2 billion years ago ...... Eukaryotic life evolves.

1.8 billion years ago .... The supercontinent Columbia forms.

1 billion years ago	The supercontinent Rodinia forms; its pieces come together to form the supercontinent Pannotia.
Less than 1 billion years ago	Multicellular life emerges.
800 million years ago	Short periods of extremely cold climates for the following 200 million years that challenge life.
540 million years ago	The Cambrian explosion; the development of hard shells and bones allows for the regular fossilization of body parts.
300 million years ago	The last great supercontinent, Pangaea, comes together.
245 million years ago	The Permian/Triassic extinction marks the end of the Paleozoic era and the start of the Mesozoic era.
200 million years ago	Pangaea starts to break up, creating the Atlantic Ocean and closing up the Tethys Sea; dinosaurs become the dominant species in many different environmental niches.
65 million years ago	The Cretaceous extinction brings about the end of the Mesozoic era and the start of the Cenozoic era.
Around 60 million years ago	The collision of India into China begins to form the Himalayan Plateau, which is the most impressive mountain range in recent geologic history; global climates continuously cool over the subsequent 60 million years.
2 million years ago	A period of extreme Ice Ages begins.

200 thousand years ago..............*Homo sapiens* evolves and becomes the dominant agent of geologic change at Earth's surface.

10 thousand years ago................The most recent warm inter-glacial period.

Geologic History

1556 ....Georgius Agricola publishes *De Re Metallica*, (*Of Things Metal*), which establishes the foundation for exploration geology.

1620 ....Francis Bacon publishes his observation that the continents could have once been attached; sea-going explorers begin to discover the outlines of the continents.

1681 ....Thomas Burnet publishes *Sacred Theory of the Earth*, in which he proposes that Earth began as a homogeneous sphere of dust and rock and then differentiated into three major layers.

1760 ....John Mitchell is the first to publish the suggestion that earthquakes might be the result of layers of rocks rubbing against each other.

1778 ....Compte de Buffon uses experiments on the cooling rate of iron to determine that Earth must be at least 75,000 years than Biblical predictions; he also suggests that climate change is the cause of changes in the appearances of organisms over time, a precursor to the ideas of natural selection and evolution.

1785	James Hutton presents the idea that geological processes have repeated themselves over and over in repeating cycles, leading him to suggest that Earth must be very old and "that we find no vestige of a beginning, no prospect of an end."
1815	William Smith publishes the first large-scale geologic map covering England and Wales, observes the correlations between fossils and geologic strata, and develops the principle of faunal succession.
1830	Sir Charles Lyell incorporates the ideas of James Hutton into the theory of Uniformitarianism, which he publishes in *Principles of Geology*; Lyell states that the geologic processes that are at work today, such as mountain building and stream erosion, operate the same way they always have in the past.
1837	Louis Agassiz publishes his proposition that northern latitudes were once covered with glaciers during an Ice Age.
1859	Charles Darwin publishes *The Origin of Species*, which shows that life is intimately dependent upon the environmental conditions within which it exists.
1869	Sir William Thomson (also known as Lord Kelvin) publishes "On Geological Dynamics" and other writings, which calculate the age of Earth as 20–100 million years old.

1898 ....Ernest Rutherford begins his work on the demonstration of radioactivity; over the next three and a half decades, many scientists (including George Darwin, John Joly, Frederick Soddy, Bertram Boltwood, and Arthur Holmes) work to develop radiometric dating, eventually determining that Earth is at least several billion years old.

1906 ....Richard Dixon Oldham discovers that Earth has a dense core, demonstrates that P, S, and surface waves can be used to determine the structure and composition of Earth's interior, and suggests that Earth's core is liquid.

1909 ....Andrija Mohorovicic uses the arrivals of refracted P and S waves to determine that there is a shallow boundary between Earth's crust and mantle, later named the Moho.

1912 ....Alfred Wegener puts forward the hypothesis of continental drift.

1928 ....Arthur Holmes lends support to the ideas of continental drift by demonstrating that convection can occur in the mantle, allowing the continents to move as if they were on a conveyor belt.

1936 ....Inge Lehmann determines from reflected P waves that the core is divided into an inner and outer core, and suggests that the inner core is solid.

1949 ....Willard Libby discovers the technique of radiocarbon dating.

1953 ....Maurice Ewing and Bruce Heezen discover a long, continuous rift running down the center of the Mid-Atlantic Ridge.

1957 ....Roger Revelle and Hans Suess demonstrate that human gas emissions are causing an increased greenhouse effect.

1960 ....Harry Hess proposes that ocean seafloor forms at mid-ocean ridges and is destroyed at ocean trenches, opening the door for the rebirth of the continental drift theory in the form of plate tectonics.

1962 ....Robert Coats publishes an article on the mechanism of the subduction of oceanic crust at island arc trenches.

1963 ....Fred Vine and Drummond Matthews show that the reversing polarity of Earth's magnetic field forms parallel bands of magnetic anomalies within the ocean crust, providing a means of remotely determining the ages of oceanic crustal rocks.

1963 ....J. Tuzo Wilson discovers hot spots.

1967 ....W. Jason Morgan and Dan McKenzie develop the details of the theory of plate tectonics from various observations made in several different fields.

1969 ....Neil Armstrong and Buzz Aldrin are the first humans to walk on the moon; analysis of moon rocks proves that Earth and the moon formed from the collision of a Mars-sized body with early Earth.

1980 .... Luis and Walter Alvarez propose that a meteorite impact caused the significant interruption in global climate that lead to the extinction of the dinosaurs 65 million years ago.

1981.. Adam Dziewonski and Don Anderson publish PREM (Preliminary Reference Earth Model), the first accurate model of the density and seismic properties of Earth, layer by layer, from the surface to the core.

1991 .... The first extra-solar planet is discovered.

1999 .... The world population reaches 6 million people.

2004 .... Mars rovers Spirit and Opportunity begin several years of exploration of the surface of Mars, proving that water once flowed on the surface of Mars and that it might therefore once have supported life.

Glossary

aquifer: Rock or soil through which groundwater moves easily.

asthenosphere: A weak layer of rock beneath the lithosphere that is close to its melting point and flows relatively easily, allowing for the motions of overlying plates.

Big Bang theory: The theory that proposes that the universe originated as a single point of energy that subsequently expanded, pulling space along with it.

biosphere: All life on Earth; the parts of the solid Earth, the hydrosphere, and the atmosphere in which living organisms can be found.

carbon cycle: The combined processes—including photosynthesis, decomposition, and respiration—by which carbon cycles between the major reservoirs of the atmosphere, the oceans, and living organisms.

chemical weathering: The processes by which the internal structure of a mineral is altered by the removal and/or addition of elements.

conduction: The transfer of heat through matter by molecular collisions; this is the dominant way heat is transferred across the core-mantle boundary and across the lithosphere.

convection: The transfer of heat through the movement of material; this is the dominant way that heat is transferred across the outer core, the mantle, the oceans, and the atmosphere.

core: The innermost layer of Earth which is primarily (85%) made of iron and is divided into a solid inner core and liquid outer core.

Coriolis effect: The apparent deflective force of Earth's rotation on all free-moving objects at Earth's surface, which creates the magnetic field in the outer core and drives fluid flow in the oceans and the atmosphere.

covalent bond: A chemical bond formed by the sharing of one or more electrons—especially pairs of electrons—between atoms.

crust: The thin, rocky outermost layer of the solid Earth; it takes the form of either continental or oceanic crust.

earthquake: The sudden slip of rock across a planar fault that releases seismic waves.

electromagnetic spectrum: The range of electromagnetic radiation according to wavelength, which includes radio waves, microwaves, infrared light, visible light, ultraviolet light, X-rays, and gamma rays.

ENSO (El Niño Southern Oscillation): The irregular cyclic swing of warm and cold phases in the tropical Pacific.

erosion: The removal and transportation of rock or soil by water, ice, wind, or gravity.

eukaryote: Complex single-celled organisms that first evolved about 2.5 billion years ago.

fault: A fracture within Earth along which movement occurs.

floodplain: The flat, low-lying portion of a stream valley that is subject to period flooding.

fold: Layer of rock that is bent, usually as the result of plate interactions.

geotherm: The vertical profile of Earth's temperature as a function of depth.

global warming: The increase in average temperature of Earth's atmosphere and surface due in part to the increase in carbon dioxide levels.

greenhouse effect: The heating of Earth's surface and atmosphere from solar radiation that is absorbed and re-emitted by the atmosphere, mainly through water vapor and carbon dioxide.

Hadley cells: Thermally-driven bands of atmospheric circulation symmetric about the equator that was first proposed by George Hadley as an explanation for the trade winds.

hot spot: A region of higher than normal volcanism at the surface that is usually associated with a concentration of heat in the mantle and thought to result from a rising plume of deep mantle rock.

hydrosphere: The water portion of Earth which is found not only at the surface but also within the atmosphere and the solid geosphere.

igneous rock: Rock formed by the crystallization of molten magma.

ion: An atom or molecule that possesses an electric charge through the gain or loss of electrons.

ionic bond: A chemical bond between two ions with opposite charges.

isostasy: The concept that Earth's crust sits at elevations that are laterally balanced gravitationally.

isotope: An atom with the same number of protons but a different number of neutrons for a given element.

Kuiper belt: A disk-shaped region in the outer solar system lying beyond the orbit of Neptune and extending to a distance of about 50 astronomical units; contains thousands of small icy bodies, some of which are on highly elliptical orbits and periodically visit the inner solar system as comets.

lithification: The conversion of newly deposited sediment into solid rock through the pressure of overlying sediment.

lithosphere: The rigid outer layer of Earth containing the crust and the uppermost mantle which is broken into Earth's tectonic plates.

magma: A body of molten rock within Earth that contains dissolved gases and may include solid crystals.

magnetosphere: A region surrounding Earth extending to several thousand kilometers above the surface where charged particles from the sun are trapped and deflected by Earth's magnetic field.

mantle: The rocky layer of Earth above the core and below the crust, containing 83% of Earth's volume and two-thirds of its mass.

mantle convection: The slow cycling movement of rock across the mantle, which is responsible for driving plate tectonics and allowing Earth to cool.

mantle plume: A rising mass of hot rock within the mantle, carrying heat from the deep mantle toward the surface and possibly supplying the source for hot spot volcanoes.

mass movement: The downslope movement of rock, regolith, and soil under the direct influence of gravity.

mechanical weathering: The physical disintegration of rock, resulting in smaller fragments that are subsequently removed as sediment.

metamorphic rock: Rock formed by the alteration of pre-existing rock through the influences of high temperatures, pressures, and/or chemically active fluids.

mid-ocean ridges: A continuous elevated zone on the floor of all major ocean basins with rifts that represent divergent plate boundaries.

Milankovitch cycles: Periodic variations in Earth's orbital parameters (orbit ellipticity, axis tilt, axis precession) that affect the distribution of solar radiation reaching Earth and causing climate changes.

moraine: A ridge of unsorted sediment deposited by a glacier.

nucleosynthesis: The process by which heavier chemical elements are synthesized from hydrogen nuclei in the interiors of stars.

Oort cloud: A swarm of comets orbiting the sun at a distance of one to two light-years, proposed as a source of some comets that pass near the sun.

orogenic belt (orogenesis): A linear region—also called orogen or fold belt—that has undergone folding or other deformation during plate collisions.

ozone layer: A region of the upper atmosphere, between about 15 and 30 kilometers in altitude, containing a relatively high concentration of ozone (a molecule of three atoms of oxygen) that partially absorbs solar ultraviolet radiation.

paleomagnetism: The permanent magnetization acquired by rock that can be used to determine the location of the ancient magnetic poles at the time it became magnetized.

Pangaea: The supercontinent that existed between 350 and 200 million years ago and broke apart to form the current configuration of continents.

permeability: The measure of a material's ability to transmit fluids.

planetesimal: One of millions of small bodies that accreted and orbited the sun during the formation of the planets.

plate: A fragment of Earth's lithosphere that moves laterally across the surface.

plate tectonics: The theory that describes how Earth's plates interact in various ways to produce earthquakes, volcanoes, mountains, and many other features of Earth's geology.

porosity: The percentage of rock or soil that consists of open spaces that could be occupied by fluids.

pressure release: The expansion of rock as the pressure is lowered; the dominant mechanism for melting beneath mid-ocean ridges.

prokaryotes: Simple one-celled organisms that first evolved on Earth more than 3.5 billion years ago.

protoplanetary disk: A rotating accretionary disk of dense gas surrounding a young, newly formed star.

pyroclastic flow: A destructive, burning, dense cloud of gas, ash, and rock that can flow down the side of a volcano during an eruption.

radiation: The transfer of energy through space in the form of electromagnetic waves; this is the way heat escapes from Earth's surface out into space and how the planet cools down over time.

radioactivity: The spontaneous decay of certain unstable atomic nuclei.

radiogenic heat: Heat released inside Earth through the radioactive decay of isotopes of potassium, uranium, and thorium; it is responsible for powering mantle convection and plate tectonics.

radiometric dating: The process of calculating the absolute ages of rocks and minerals by counting the atoms of radioactive isotopes and their byproducts.

rheology: The study of the deformation and flow of matter.

Ring of Fire: An extensive zone of volcanic and seismic activity that coincides roughly with the borders of the Pacific Ocean and is due to the large numbers of subduction zones present there.

rock cycle: The interrelated sequence of events by which rocks are initially formed, altered, destroyed, and reformed as a result of magmatism, crystallization, erosion, sedimentation, metamorphism, and melting.

sediment: Loose particles—created by the weathering and erosion of rock, chemical precipitation from aqueous solution, or from secretions of organisms—which are transported by water, wind, or ice.

sedimentary rock: Rock formed from the lithification of sediment that has been eroded, transported, deposited, compacted, and cemented.

seismic tomography: A method of imaging Earth's interior through the analysis of the paths of seismic waves from earthquakes recorded on seismometers around the world.

shield volcano: A broad, gently sloping volcano built from successive flows of fluid basaltic lavas.

silicate mineral: A mineral formed from the bonding together of silicon/oxygen tetrahedra, usually in combination with other elements. Most minerals, and therefore most rocks, are silicates.

slab pull: The force of gravity acting upon the sheet of subducting oceanic lithosphere, which is the dominant force driving the particular motions of plate tectonics.

solidus: The temperature, at a particular pressure corresponding to a depth within Earth, at which a mineral begins to melt.

stratosphere: The layer of the atmosphere above the troposphere that contains the ozone layer.

stratovolcano: A volcano composed of both lava flows and ashfalls; also called a composite cone volcano.

supercontinent: A continental land mass containing most if not all of the continents; there have been several over the course of Earth's history.

supernova: An exploding star of intense brightness; it is during this final phase of a star's lifetime that elements heavier than hydrogen and helium are formed.

troposphere: The lowermost layer of the atmosphere, which contains Earth's clouds and weather.

unconformity: A boundary within rock layers that represents a break in the rock record, caused by the erosion or the lack of deposition of rock.

volcanic island arc: A chain of volcanic islands generally located about 100 kilometers above a subducting slab of oceanic lithosphere; the magma forms from the entry of subducted ocean water into the mantle.

water cycle: The constant movement of water among the oceans, atmosphere, geosphere, and biosphere.

water table: The upper boundary of the saturated zone of groundwater.

weathering: The disintegration and decomposition of rock at or near Earth's surface.

Wilson cycle: The repeating process, first identified by J. Tuzo Wilson, by which oceans open and close, and supercontinents form and break apart.

Biographical Notes

Louis Agassiz (1807–1873): Swiss-American zoologist and geologist best known for discovering the existence of previous Ice Ages. Trained first as a medical doctor, Agassiz learned zoology from Cuvier and geology from von Humboldt and made great contributions in both areas. Most of Agassiz's career was spent studying fish, but he also spent time studying the glaciers of Switzerland and later figured out that all places that showed evidence of large amounts of loose, unsorted glacial debris had once been under ice (as Greenland was currently). Agassiz later came to America as a professor at Harvard and was revered for many years as America's greatest scientist. Agassiz was one of the last prominent zoologists to reject Darwin and his ideas on evolution, based on religions grounds. He also believed that blacks were inferior to whites, which was used by slave owners as a justification for slavery.

Georgius Agricola (1494–1555): German scientist who laid the foundation of mining and mineralogy through the publication of several famous books. He spent much of his life as a physician but was attracted to geology and laid the foundations for physical geology in 1544 with the book *De Ortu et Causis Subterraneorum*. This was followed in rapid succession over the next six years with books that defined and described the discovery of minerals and ores. His most famous book, *De re Metallica*, set the standards for mining and metallurgy for the next several centuries. It was so popular that it was even republished in 1912 in *Mining Magazine* with an English translation by the American mining engineer Herbert Hoover (better known as the 31st president of the United States!) and his wife, Lou Hoover.

Svante Arrhenius (1859–1927): Swedish chemist who is one of the founders of the field of physical chemistry, which governs the way chemical reactions occur. He was the first to quantify the amount of heat that was needed to be added to chemical materials for a reaction to occur, called the activation energy; the equation describing this relationship is called the Arrhenius equation in his honor. After receiving a Nobel Prize for his physical chemistry work, Arrhenius turned his attention to geology, making contributions to the fields of astronomy, astrophysics, and cosmology. His most significant geological contribution, however, was in predicting the "greenhouse effect" that would be caused by the increased production of carbon dioxide. Arrhenius predicted that a doubling of atmospheric carbon would increase global temperatures by 5°C–6°C, which is remarkably close to recent estimates by the IPCC of 2°C–4.5°C. Because he thought that warmer climates would mean better agriculture, however, he advocated increased CO_2 production.

A. Francis Birch (1903–1992): American geophysicist who intimately understood geophysics at both large scales (seismic waves, mantle structure) and small scales (atomic interactions), and combined both into the first accurate assessment of Earth's interior composition. Birch used observations of seismic wave velocities and laboratory experiments on minerals to approximate the compositions of the mantle and core. He figured out how to extrapolate surface measurements to deep-Earth conditions and how to identify minerals by their seismic properties. Though considered a serious man, Birch allowed himself a bit of humor in a now-famous footnote to his epic (both in importance and length) 1952 paper *Elasticity and Constitution of the Earth's Interior*, where he provided the following sagacious warning to all who study Earth's interior:

> Unwary readers should take warning that ordinary language undergoes modification to a high-pressure form when applied to the interior of the Earth. A few examples of equivalents follow:

High Pressure Form	*Ordinary Meaning*
Certain	Dubious
Undoubtedly	Perhaps
Positive proof	Vague suggestion
Unanswerable argument	Trivial objection
Pure iron	Uncertain mixture of all the elements

Wallace Broecker (b. 1931): American geochemist who laid the foundation for future research in the connections between oceans and climate and the cycling of carbon between different reservoirs. He revealed the intricacies between different environmental systems and used radioisotopes to examine climate change records. Broecker is best known for showing the significant role that ocean circulation plays in changing atmospheric climate and causing abrupt climate change. His research has established the standards for chemical oceanography.

Thomas Burnet (1635–1715): English theologian, cosmogonist, and Royal Chaplain to King William III who published his speculative account of the past formation and future of Earth in the *Telluris Theoria Sacra*, or *Sacred Theory of the Earth*. In it, he described how Earth began as a homogeneous sphere, became layered (the middle layer being water), broken (with the water coming to the surface as Noah's flood), and will eventually become a star during Armageddon. The broken outlines of the continents, he asserted, were the result of Noah's catastrophe. This model was almost entirely based upon Christian theological grounds. Although Burnet dabbled at science (he made sounding measurements of shallow ocean regions to try to estimate the amount of water on Earth), the book is written in a scientific style. Burnet was skeptical in his religion, however, and was forced to leave his court position when he suggested, in the 1692 book *Archaeologiae Philosophicae*, that the six days of Genesis and the Fall of Man were symbolic rather than literal events.

Georges Cuvier (1769–1832): French naturalist and zoologist. No person knew more about animals in his era than Cuvier did, and this had good and bad consequences. Because of Cuvier's reputation and stature (he managed to maintain his position as top scientist in France before, during, and after Napoleon's rule), he was able to gain acceptance of his revolutionary view that most fossils were the remains of organisms that had since gone extinct. Before then, people thought that God would not allow any of his creatures to go extinct, so animals found only as fossils must still be living somewhere around the globe. Cuvier's establishment of extinctions thus set the stage for geologists to determine relative ages of rocks based upon index fossils. Cuvier became a champion of catastrophist theories, proposing that extinct animals must have died in a set of catastrophic events. Catastrophism fell out of favor with the public acclaim of Hutton and Lyell, but, interestingly, with the 20th-century discovery of mass extinctions, Cuvier's views on Catastrophism have been seen in a more positive light. Where Cuvier's influence did harm, however, was in his rejection of the theory of the gradual evolution of species. So sure was Cuvier of the sudden extinction of species that he ridiculed ideas of gradual evolution, and this kept many biologists from considering the ideas of evolution until Darwin published *Origin of the Species* decades after Cuvier's death.

James Dwight Dana (1813–1895): An American, Dana was the most esteemed geologist in the last half of the 19th century, making important contributions concerning volcanoes, mountain-building, and the structure and origin of continents. He was an expert on California geology and provided a great deal of information for would-be gold miners during and after the 1849 Gold Rush. He succeeded the esteemed Professor Benjamin Silliman (who was the first person to distill petroleum) on the geology faculty at Yale University, going so far as to marry his daughter. Dana was an influential writer, and his textbooks on mineralogy are still in use and have been the definitive texts for centuries: *System of Mineralogy*, first written in 1837, had its 8th edition published in 1997, and *Manual of Mineralogy*, first written in 1848, had its 22nd edition published in 2002.

Charles Darwin (1809–1882): English naturalist who was the most influential scientist of the 19th century; his 1859 book *On the Origin of Species* is one of the most influential books ever written. Darwin did not invent the ideas of evolution, which had been around for a while and furthered by scientists like Jean-Baptiste Lamarck, but developed them into a remarkably coherent science. Because of criticism of evolutionary ideas, Darwin waited two decades to publish *On the Origin of Species*. He did so only when Alfred Russel Wallace had arrived at a similar theory and the two published their ideas together in 1858. Darwin was heavily influenced by the Uniformitarian ideas of Lyell as well as his own geologic and paleontologic discoveries as a ship naturalist on the five-year voyage of the HMS *Beagle*. Darwin actually made several important geologic discoveries himself, including the origin of coral atolls, which he published in *The Voyage of the Beagle*. Darwin's theories on evolution and natural selection attracted as much controversy then as they do now but were quickly accepted by the entire scientific community, and he was awarded the Copley Medal of the Royal Society of London in 1864. His ideas were so well thought out that common views of the mechanisms of natural selection have changed only slightly over the past 150 years.

Adam Dziewonski (b. 1936): Polish-American seismologist who was one of the most influential geophysicists in modern times.

Dziewonski was one of the founders of the field of seismic tomography, which is the primary means of visualizing the interior of Earth. He showed that subducted lithosphere sank to the base of the mantle and was part of a mantle-wide convection cycle. Dziewonski also pioneered the global categorization of the magnitudes and geographical orientations of all large earthquakes, which has been the foundation for studying and understanding the regional tectonics and interactions between tectonic plates. In 1981, together with Don Anderson, Dziewonski created the best model of Earth's interior, PREM (Preliminary Reference Earth Model). Incorporating many different kinds of data, PREM quantified the density and seismic velocities at all depths within Earth, and—although more than 25 years old—is still the leading global vertical model of the planet's interior.

Walter Elsasser (1904–1991): German-American physicist considered to be the father of geodynamo theory. Elsasser explained how Earth's magnetic field was generated by eddy currents within the liquid iron outer core. Much of his work was done during and after World War II, partly in his spare time while working with the U.S. Signal Corps. Elsasser was a brilliant physicist who came close to winning the Nobel Prize twice; two of his lines of research, concerning the wave aspect of electrons and the binding energies of protons and neutrons in heavy radioactive nuclei, were carried further by other researchers who later received the Nobel Prize.

W. Maurice Ewing (1906–1974): American geophysicist and oceanographer who was one of the most influential investigators of the structure of the ocean and ocean crust, leading more than 50 oceanic expeditions. He is best known for discovering the SOFAR channel, a shallow ocean layer that efficiently carries sound throughout the oceans, and the Mid-Atlantic Rift Zone, which he discovered in 1953 with Bruce Heezen. Ewing was the founder and first director of the Lamont-Doherty Earth Observatory at Columbia University, one of the leading programs of ocean investigations.

Benjamin Franklin (1706–1790): American statesman, inventor, scientist, publisher, and writer who—in addition to everything else this amazing man did—was a pioneer in meteorology and climate change. Franklin showed that lightning was natural electricity and published many other observations of atmospheric phenomena including storms and waterspouts. In 1768 he made the first map of the Gulf Stream. He came remarkably close to predicting plate tectonics, nearly 200 years in advance, when he wrote in 1782 that "The crust of the Earth must be a shell floating on a fluid interior.... Thus the surface of the globe would be capable of being broken and distorted by the violent movements of the fluids on which it rested." In addition, Franklin observed the correlation between eruptions in Iceland in the mid-1780s and the cold weather that followed in the succeeding years. This led to his founding the field of climate change as a result of volcanic eruptions.

Galileo Galileii (1564–1642): Italian physicist who is the "father of modern astronomy" and the first to use a telescope to make inferences about the nature of the solar system. As a result of his discovery that there were four moons that orbited Jupiter, Galileo determined that Earth and the rest of the planets must be orbiting the much larger sun. Galileo used telescopes for many other studies, including charting sunspots, measuring the heights of mountains on the moon, and determining that the Milky Way was not a cloud but a mass of distant stars. Galileo also made tremendous contributions to physics and mathematics and was called the "father of modern science" by Albert Einstein.

Stephen Jay Gould (1942–2002): American paleontologist, evolutionary biologist, and historian of science who, through his colorful and clever essays that ran monthly for years in *Natural History* magazine, taught countless millions of people about paleontology, geology, and evolution. As a scientist, Gould made great contributions in the area of evolutionary biology, advocating the rapid emergence of new species through a process he called "punctuated equilibrium." It was through his writings, collected into many popular books, however, that he was best known, and for many years he was one of the most vocal and staunchest advocates for evolution and the scientific method. Gould was also a renowned baseball fan and was featured in the famous Ken Burns documentary, *Baseball*.

Beno Gutenberg (1889–1960): A German-American seismologist, Gutenberg was born and went to school in Germany and began his career as a professor there. Due to growing anti-Semitic sentiments, however, he was not able to continue his work in Germany and came to America in 1930 where, together with colleague Charles Richter, he turned the California Institute of Technology into the world's leading seismological research institute. Gutenberg was a pioneer in the field of using seismic waves to determine the structure of Earth's interior. The core-mantle boundary, which he was the first to accurately locate, is still often referred to as the Gutenberg discontinuity. Gutenberg also worked with Richter in developing the first earthquake magnitude scales, and together they also developed the Gutenberg-Richter law, which established the probability distribution of earthquakes at different energy levels.

Edmund Halley (1656–1742): English astronomer, geophysicist, mathematician, meteorologist, and physicist, who is best known for the comet named in his honor (because he predicted the year of its next return to the inner solar system). He also made contributions in a wide variety of other earth and planetary sciences. He published works about trade winds and monsoons and was the first to identify solar heating as the cause of atmospheric convection and to document the correlation between barometric pressure and the height of the sea surface. He invented a diving bell and explored the bottom of the Thames River. He even spent two years sailing in the Atlantic Ocean, documenting variations in Earth's magnetic field. One idea he got very wrong was his proposition that Earth was partially hollow, consisting of internal concentric shells separated by atmospheres. This incorrect hypothesis persisted for years and appeared in stories like Jules Verne's *Journey to the Center of the Earth* and Edgar Rice Burroughs's *Tarzan at the Earth's Core*.

Harry Hess (1906–1969): Sometimes chance plays an important role in scientific discovery. American geologist Harry Hess's research was interrupted by World War II, where he served as the captain of the USS *Cape Johnson*; the transport ship, however, was equipped with newly-developed sonar equipment, and Hess used it to map out uncharted regions of the ocean seafloor. He returned to Princeton University after the war and developed his ideas, making some of the most important observations that led to the theory of plate tectonics. His 1962 paper "History of Ocean Basins," which described how ocean seafloor moved away from mid-ocean ridges toward ocean trenches, was for a while the single most cited reference in solid Earth geophysics. Hess made many other discoveries of the ocean seafloor, including the flat-topped underwater seamounts he called guyots, which proved that ocean seafloor sank with age (because the seamounts were once above-surface islands, which is when the tops were eroded and flattened).

Arthur Holmes (1890–1965): British geologist who was involved with the two greatest geologic controversies of his time—he not only made great contributions to them but proved to be right on both counts! Holmes began using radiometric dating of rocks while an undergraduate in 1910 and made the first uranium-lead calculations (on a 370-million-year-old rock from Norway). He strongly advocated an age of Earth in the range of billions of years, decades before such ideas were widely accepted; in fact, his estimate of Earth's age of 4.5 billion years was remarkably close to being correct. Even more importantly, however, Holmes was one of the strongest advocates for continental drift at a time when the idea was widely dismissed. Holmes demonstrated that though the mantle was solid rock it should still be convecting, providing the means of moving continents laterally across Earth's surface.

Alexander von Humboldt (1769–1859): Prussian naturalist and explorer who, like some other geniuses of previous centuries, made significant contributions to many scientific fields; he made the most impact in essentially starting the field of biogeography. Von Humboldt made a long expeditionary journey to South and Central America between 1799 and 1804 and made many important discoveries about the dependence of life upon the physical environments within which it exists. Von Humboldt also made important observations about Earth's magnetic field, meteorology, and the composition of the atmosphere, and even concluded, correctly, that South America and Africa had once been connected.

James Hutton (1726–1797): Few individuals did more to single-handedly change the view of Earth than this Scottish geologist who has aptly been referred to as the "father of modern geology" on many occasions. Hutton determined that Earth's interior was hot, and that this heat converted layers of sedimentary rock into new rock, presenting the first reasonable assessment of the rock cycle. He was the founder of a new school of thought called Plutonism, which described Earth's surface as being subjected to cycles of erosion, deposition, and lithification, creating layers upon layers of rock such as his now-famous "Hutton's unconformity" in Jedburgh, Scotland; this view overturned the Neptunist views presented by Abraham Werner. In addition, Hutton demonstrated that Earth must be very old, and initiated the school of thought called Uniformitarianism. This view ran counter to the biblically based ideas of Catastrophism, which advocated an Earth only thousands of years old. Hutton's big problem was that he was a terrible writer and very verbose; *An Investigation of the Principles of Knowledge and of the Progress of Reason, from Sense to Science and Philosophy*, was 2,138 pages long, so his ideas did not take hold until later, when they were put forth by more literarily capable champions like Charles Lyell.

Thomas Huxley (1825–1895): English biologist who often appeared in the shadow of his friend and colleague, Charles Darwin (and has been referred to as "Darwin's Bulldog"), Huxley was a brilliant anatomist and contributor to the ideas of evolution in his own right. He was initially skeptical about some of Darwin's ideas, such as natural selection and gradual evolution, but became one of Darwin's most powerful advocates and supporters. Huxley himself, an expert on vertebrate anatomy, proposed that humans and apes were similar enough to have a common ancestor and also demonstrated in 1870 that birds had evolved from dinosaurs, which was an idea that did not gain acceptance until 100 years later. Huxley was also vocal on the inadequacy of using religious ideas as the basis of scientific discovery and coined the term "Agnosticism" to describe his own beliefs.

Harold Jeffreys (1891–1989): English mathematician, geophysicist, and astronomer who made contributions to the understanding of Earth's deep interior. His study of seismology helped prove in 1926 that Earth's outer core was liquid. In 1940, Jeffreys, along with geophysicist Keith Bullen, presented a remarkably accurate depth profile of the seismic characteristics of Earth from the surface to the center. Given that they had limited seismic coverage of Earth's surface and no computers to work with, it was an impressive feat. For more than 40 years this "Jeffreys-Bullen model" remained the definitive description of Earth's interior. Despite his deep insights into math and geophysics, however, Jeffreys never accepted the theory of plate tectonics and was an opponent of it until his death.

William Kaula (1926–2000): American geophysicist who made fundamental contributions to two different fields: satellite-based geodesy and comparative planetology. His geodetic work began in the military when he realized that satellites could be used for measuring Earth's surface as well as tracking missiles. His work was vital to the eventual development of the Global Positioning System (GPS) network. Kaula applied the new possibilities of geodesy to satellite studies of other planets and provided the basis for much of our understanding of how planets form and evolve over time.

Wladimir Köppen (1846–1940): Russian-German geographer, meteorologist, climatologist, and botanist. Publishing more than 500 papers in many different fields, Köppen was one of the last of the broad-discipline scientists who made contributions in many different areas. Köppen is best known for his establishment of a classification of climates (the Köppen Classification) which is still used today. He spent a good deal of his career refining this classification but also did important work with his son-in-law Alfred Wegener in providing crucial evidence to support the Milankovitch theory of the cause of Ice Ages.

Inge Lehmann (1888–1993): Danish seismologist who, in 1936, cleverly demonstrated that Earth had a solid iron inner core using seismic P waves that traveled through the inner core, 30 years before seismologic studies that used modern seismographs. Lehmann made contributions to seismology for many decades but did not overstate her findings; her seminal 1936 paper on the inner core was titled P' (the name of the seismic phase used in the findings).

Mikhail Vasilievich Lomonosov (1711–1765): Russian scientist and writer whose legendary activities span nearly all areas of human activity and research. One of the most accomplished individuals of the 18th century, his research provided groundwork in the fields of thermodynamics, gravity, the wave property of light, material phase properties, and the theory of gases. He observed Venus's atmosphere, explained the formation of icebergs, determined the organic origin of coal and petroleum, and cataloged more than 3,000 minerals. In addition, he reformed the Russian language, wrote poems, and was an artist who set up the first stained-glass mosaics outside of Italy.

Charles Lyell (1797–1875): The histories of Lyell and James Hutton are forever interlinked. A Scottish geologist, Lyell made significant contributions to the field. He was an excellent writer, however, and his textbooks *Principles of Geology* (1830) and *Elements of Geology* (1838) were the most widely read and influential books on geology in the 19th century. It was in this role of educator that Lyell became an advocate for the earlier work of Hutton and a champion of the philosophy of Uniformitarianism. Lyell wrote that "the present is the key to the past," meaning that the geologic processes that have shaped our world are the same ones in operation today; this suggested that Earth was immensely old. Lyell was also a friend of Charles Darwin and a supporter of his theories on evolution.

Milutin Milankovitch (1879–1958): Serbian civil engineer and geophysicist who, though hindered by both world wars, managed to discover and publicize the connection between fluctuations in Earth's orbital parameters (orbit ellipticity, axis tilt, and axis precession), the distribution of solar radiation, and global climates—providing the explanation for why the recent Ice Ages had occurred. This work was preceded by his computation of the first accurate curve of solar insolation on Earth's surface.

W. Jason Morgan (b. 1935): An American geophysicist, Morgan was a young professor at Princeton University when he published one of the most influential papers ever in the history of earth science: the 1968 paper "Rises, Trenches, Great Faults, and Crustal Blocks," which established the fundamentals of the theory of plate tectonics. The evidence supporting plate tectonics had been around for many years, but it took a totally new way of thinking to put the pieces together, and Morgan was the one to do it (along with Dan McKenzie, working independently in England). Morgan went on a few years later to further the ideas of J. Tuzo Wilson about mantle hot spot plumes, thus providing the means of determining the directions and velocities of the moving tectonic plates. Morgan did not write many papers over his career, but many of the ones he wrote were very influential, and most of Earth science of the past 40 years is built upon his 1968 paper.

Hans Oeschger (1927–1998): German climatologist who revolutionized our ability to document climate change and was one of the first scientists to prove the correlation between temperatures and carbon dioxide concentrations in the atmosphere. Oeschger developed a way to use radioactive isotopes to determine the ages of water in the deep parts of the oceans, quantifying the rates of ocean circulation. He then applied his methods to glacial ice and was able to document atmospheric compositions going back 150,000 years. He found that carbon dioxide levels were 50% lower during Ice Ages than they currently are.

Roger Revelle (1909–1991): American oceanographer who was instrumental in understanding the connection between the compositions of the ocean and the atmosphere. He showed that the oceans would not be able to absorb carbon dioxide at the rate that it was being added to the atmosphere from anthropogenic sources. He was one of the first scientists to accurately quantify increasing atmospheric CO_2 levels and show how the "greenhouse effect" was leading to global warming.

Charles Richter (1900–1985): Before this American seismologist began his studies of the sizes and energies of seismic waves, there was no accurate way to assess the magnitudes of earthquakes. In 1935, working with fellow CalTech professor Beno Gutenberg, Richter developed a logarithmic scale that incorporated both the amplitudes of waves on seismograms with the distance from the earthquake in order to provide an accurate measure of the size of an earthquake. Richter and Gutenberg later went on to relate these earthquake magnitudes to the actual levels of energy released. Richter also published the first major textbook on seismology, *Elementary Seismology*.

A. E. Ringwood (1930–1993): Australian geologist who had an uncanny ability to interpret the composition of Earth's deep mantle based upon a small number of very difficult high-pressure mineral physics experiments. Ringwood was the first person to accurately determine the composition of the mantle, demonstrating how olivine and pyroxene undergo mineral phase changes at depths of 410 and 660 kilometers. He accurately predicted that the upper and lower mantles had mostly the same composition, with most differences in behavior due to the different phases at different pressures. Ringwood was also one of the first people to understand the effects of mantle composition on plate tectonics and the subduction of slabs of ocean lithosphere.

William Smith (1769–1839): English geologist who invented the geologic map through his creation the first geologic map of England. Smith noticed that layers of rock could be identified over large distances and used color coding to map them across the country. Smith also developed the principle of faunal succession, which allowed him to use different fossils as identifying markers for different strata. Unfortunately for Smith—who began his career as a poor surveyor's assistant—his work was never accepted or appreciated by the upper-class geologic establishment and was often plagiarized; he died in debtor's prison.

Nicolaus Steno (Niels Steensen) (1638–1686): Danish anatomist and geologist who established the fundamental principles of stratigraphy, which may seem obvious to us now but were breakthroughs in understanding geology. These fundamentals were the law of superposition (younger layers are deposited on top of older layers), the principle of original horizontality (layers are initially deposited as horizontal sheets), the principle of lateral continuity (layers extend over broad areas), and the principle of cross-cutting (a layer cutting across another layer must be younger). Steno was also one of the earliest paleontologists, observing the similarities between fossils and living organisms.

Harold Urey (1893–1981): An American physical chemist, Urey began his career working with Niels Bohr on atomic structure, receiving a Nobel Prize for his discovery of deuterium. He went on, however, to establish the field of cosmochemistry and made groundbreaking discoveries about the early composition of Earth and the solar system. Urey made the first estimates of the composition of Earth's early atmosphere and with one of his graduate students, Stanley Miller, made inspiring experiments that showed that amino acids, the building blocks of life, would naturally form from simple components. Urey was also a leader in the Manhattan Project during World War II, creating the fissionable uranium-235 used for the nuclear bombs.

James Van Allen (1914–2006): American space scientist who was the pioneer of space physics, using satellites and balloons to increase the world's knowledge of energetic particles, plasmas, and radio waves throughout the solar system. Van Allen began his career with the military and was instrumental in getting satellites up into space, playing a key role in the Cold War space race. After the war, Van Allen and his research program made groundbreaking observations of the upper atmospheres of Earth, Venus, Mars, Jupiter, Saturn, Uranus, and Neptune. Earth's high-atmosphere Van Allen Radiation Belts are named after him.

Charles Walcott (1850–1927): An esteemed American paleontologist who is best known for his discovery of the Cambrian fossils of the Burgess Shale in Alberta, Canada. Walcott had specialized in Cambrian period fossils and identified the most important fossil find in history, revealing the sudden explosion of life forms that occurred 540 million years ago. Walcott was also one of the most influential scientific leaders in the country, directing, at different times, the United States Geologic Survey, the American Association for the Advancement of Science, and the Smithsonian Institution, as well as helping to found the Carnegie Institute of Washington.

Alfred Russel Wallace (1823–1913): English naturalist, explorer, geographer, anthropologist, and biologist who co-developed the ideas of natural selection, simultaneously and in competition, with Charles Darwin. Wallace was a world expert on the distribution of species in different geographic regions and is considered the "father of bioecology." He was a developer of the ideas of evolution, and also pioneered several environmental fields such as concerns over human impacts on environments, deforestation, and the invasion of foreign species. Public opinion of Wallace was mixed, largely because of his support of several pseudo-spiritual movements such as phrenology and mesmerism, but he is now recognized as one of the most influential scientists establishing the connection between evolution and geographic environments.

Alfred Wegener (1880–1930): A meteorologist, Wegener was trained in astronomy and did most of his research in meteorology, but is known as the author of the theory of continental drift. Wegener traveled on expeditions to Greenland and pioneered the use of balloons to track atmospheric air currents, but became fascinated by the geologic and geographic evidence that the continents were once connected. He put forth his ideas in 1912 and published them in 1915 as *The Origin of Continents and Oceans*. Partly because of his lack of background in geology and partly because the mechanisms he proposed for the continental motions were preposterously unreasonable, his ideas were never broadly accepted within the scientific community. He was right, of course: The continents *do* move and *were* once connected in a supercontinent that he named Pangaea, but only a handful of prominent scientists during his day had the insight and bravery to publicly support him.

Abraham Werner (1749–1817): A German geologist, Werner did not publish much but was renowned as a brilliant teacher and took on a large following of supporters. Werner became the champion of a philosophy of Earth's formation called Neptunism, whereby all of Earth's surface rocks precipitated out of a giant ocean that once entirely covered Earth's surface. This form of Catastrophism, with Earth's rocks forming quickly out of a single geologic event, was in direct conflict with the Uniformitarianism of Hutton and Lyell. Werner received support from biblical supporters because they reconciled his giant ocean with Noah's flood. Werner was not able to reconcile the many inconsistencies of his theory (such as how volcanic rocks like basalt formed) and it eventually fell out of favor, but he was the most influential geologist of his day.

J. Tuzo Wilson (1908–1993): A Canadian geophysicist who, except for the vagaries of publishing, might well be credited as the discoverer of plate tectonics. Several years before Morgan and McKenzie are credited with establishing plate tectonics, J. Tuzo Wilson described how the lithosphere moved over a weaker asthenosphere, and how the Hawaiian Islands formed when the Pacific Plate moved over a mantle hot spot. He also was the first person to accurately describe the mechanisms of transform faults like the San Andreas Fault. Paper reviewers, however, rejected his ideas, and he ended up publishing in a Canadian journal with a small circulation. The Wilson cycle, which describes the process by which oceans open and close over long periods due to plate motions, is named after him.

Bibliography

Alley, Richard. *The Two-Mile Time Machine: Ice Cores, Abrupt Climate Change, and Our Future.* Princeton: Princeton University Press, 2002. Ice cores have provided some of the best records of climate change, and a pioneer in the field explains the science behind them and their implications for future climates.

Atwater, Brian. *The Orphan Tsunami of 1700.* Seattle: University of Washington Press, 2005. A description of the stunning discovery of the enormous magnitude 9.0 earthquake that occurred in the Pacific Northwest in 1700.

Aubrecht, Gordon. *Energy: Physical, Environmental and Social Impact.* Boston: Benjamin Cummings, 2005. An excellent text that describes the wide variety of energy sources, including their trends, long-term prospects, and resource supplies.

Benn, Douglas I., and David Evans. *Glaciers and Glaciation.* London: Edward Arnold Publishers, 1998. A premier guide on glaciology, this book looks at the geology and physics of glaciers, the flow of ice, and the erosion that results.

Berner, Elizabeth K., and Robert Berner, *Global Environment: Water, Air, and Geochemical Cycles.* Upper Saddle River, NJ: Prentice Hall, 1996. Requiring some basic understanding of chemistry, this book provides a good overview of global geochemistry and environmental problems and establishes the fundamental concepts of geology, biogeochemistry, oceanography, meteorology, and limnology.

Bjornerud, Marcia. *Reading the Rocks: The Autobiography of the Earth.* New York: Westview Press, 2005. A nicely written overview of the earth, from its formation to projections of its future.

Bolt, Bruce. *Earthquakes.* New York: W. H. Freeman & Company, 2003. The classic standard introductory discussion of earthquakes and seismology.

Bowen, Mark. *Thin Ice: Unlocking the Secrets of Climate in the World's Highest Mountains.* New York: Holt Paperbacks, 2006. Scientist and expert climber Mark Bowen joins climatologist Lonnie Thompson and his crew on the highest and most remote glaciers along the equator to study the history of climate change.

Breining, Greg. *Super Volcano: The Ticking Time Bomb Beneath Yellowstone National Park*. Osceola, WI: Voyageur Press, 2007. An examination of the volcanic hazard posed by the Yellowstone hot spot that describes the possible results of another enormous eruption in this region.

Bridgeman, Howard A. *Global Climate System*. Cambridge: Cambridge University Press, 2006. A valuable assessment of the world's climate diversity, with each chapter containing an essay by a specialist in that field.

Broecker, Wallace. "Glaciers that Speak in Tongues and Other Tales of Global Warming." *Natural History*, 110 (October 2001): 60–70. One of the world's most influential climatologists gives his account of the climate change and the science behind it.

Brown, Geoffrey C., and Alan Mussett. *The Inaccessible Earth*. London: Chapman and Hall, 1993. An advanced college text that sets the standard for discussions on the composition of Earth's crust and mantle.

Brumbaugh, David. *Earthquakes, Science and Society*. Upper Saddle River, NJ: Prentice Hall, 1998. A very nice and concise overview of the many impacts of earthquakes on humanity.

Bryson, Bill. *A Short History of Nearly Everything*. New York: Broadway Books, 2004. Imagine trying to describe the history of the universe and its discovery (incorporating physics, chemistry, biology, and geology) with engaging and casual prose and succeeding—that is this book.

Bullard, Fred Mason. *Volcanoes of the Earth*. Austin: University of Texas Press, May 1984. This book traces the growth of volcanology with discussions of geothermal energy, the environmental effect of volcanoes on climate, air and soil pollution, and the cyclic nature of volcanic eruptions.

Calvino, Italo. *Cosmicomics*. Orlando, FL: Harvest Books, 1976. A brilliant work of fiction that provides incredible insights to what it might have been like to exist through the creation and formation of our universe. (The essay "All at One Point," which describes hypothetical "life" before the Big Bang in 0 dimensions, is my favorite!)

Carson, Rachel. *The Sea Around Us.* New York: Oxford University, 1951. One of the most influential and widely-read books in the history of science. It is beautifully and simply written.

Clarke, Robin, and Jannet King, *The Atlas of Water: Mapping the World's Most Critical Resource.* New York: New Press, 2004. A good account of the many complex issues surrounding one of Earth's most important and threatened natural resources.

Cone, Joseph. *Fire Under the Sea.* New York: Quill/William Morrow, 1992. The story of the discovery of volcanoes and thermal vents at mid-ocean ridges, and the strange environments that they support.

Cousteau, Jacques-Yves. *The Silent World.* New York: Harper and Row, 1953. The famous memoir by one of the most influential explorers of the unusual worlds beneath the ocean surface.

Cox, John D. *Climate Crash.* Washington, D.C.: Joseph Henry Publisher, 2005. An excellent presentation in very readable language of the many aspects of the complex systems of climate change and the implications for humanity.

Davidson, Jon P., Walter Reed, and Paul Davis. *Exploring Earth: An Introduction to Physical Geology, 2nd ed.,* Upper Saddle River, NJ: Prentice Hall, 2002. An introductory college physical geology text that does a very good job at showing the different components of the rock cycle from a process-oriented viewpoint.

Davies, Geoffrey F. *Dynamic Earth: Plates, Plumes and Mantle Convection,* Cambridge: Cambridge University Press, 1999. An advanced, but mostly qualitative, discussion of the dynamics of mantle convection by one of its founding discoverers.

Davies, Paul. *The Goldilocks Enigma: Why Is the Universe Just Right for Life?* Boston: Mariner Books, 2008. An entertaining book on the multiverse and other surprising theories put forth to answer the question of existence, offering both descriptions of the science behind the theories and the philosophical implications.

De Boer, Jelle Zeilinga, and Donald Sanders. *Earthquakes in Human History.* Princeton, NJ: Princeton University Press, 2004. A good overview of the societal effects of many large historical earthquakes.

———. *Volcanoes in Human History.* Princeton, NJ: Princeton University Press, 2004. Volcanoes have had significant effects on global climates, and the accounts of many such eruptions are presented here.

Decker, Robert, and Barbara Decker. *Volcanoes, 4th ed.* New York: W. H. Freeman, 2005. The classic introductory text to volcanoes and volcanology, widely used in introductory classes throughout the country.

————. *Volcanoes in America's National Parks.* New York: Odyssey Publications, 2001. This book not only talks about the volcanoes that are part of 31 different national parks and monuments but gives travel information on how to actually go and see them for yourselves.

Deffeys, Kenneth. *Beyond Oil: The View from Hubbert's Peak.* New York: Hill and Wang, 2006. An influential assessment by an authoritative geologist on the production futures of petroleum and the implications for the future of energy use.

Diamond, Jared. *Collapse: How Societies Choose to Fail or Succeed.* New York: Penguin, 2005. A dense comparative study of societies that have sometimes fatally undermined their own ecological foundations.

————. *Guns, Germs, and Steel: The Fates of Human Societies.* New York: W. W. Norton & Company, 1999. A brilliant treatise, a bit wordy at times, on the influence of geography and geology on the course of human civilization.

Dietrich, Richard V., and Brian Skinner, *Rocks and Rock Minerals.* New York: John Wiley & Sons, 1979. A visually appealing guide to rocks, explaining the different ways that they form and giving many attractive examples of them.

Emanuel, Kerry. *Divine Wind: The History and Science of Hurricanes.* New York: Oxford University Press, 2005. A good description of heavy weather and storms, including accounts of many important and significant historic hurricanes.

Emiliani, Cesare. *Planet Earth: Cosmology, Geology, and the Evolution of Life and Environment.* Cambridge: Cambridge University Press, 2007. A good overview of the integrated physical and geological systems operating on our planet.

Ferris, Timothy. *The Whole Shebang: A State-of-the-Universe Report.* New York: Simon and Schuster, 1997. Ferris is one of the country's leading science writers, and here he explains the evolution of the universe in a very readable and accessible way.

Fisher, Richard V., Grant Heiken, and Jeffrey Hulen, *Volcanoes: Crucibles of Change*. Princeton, NJ: Princeton University Press, 1998. A description of the different kinds of volcanoes that are found around the world, including firsthand accounts of many eruptions.

Flannery, Tim. *The Weather Makers: How Man Is Changing the Climate and What It Means for Life on Earth*. New York: Grove/Atlantic, 2007. An up-to-date presentation of the ways that human activity continues to alter global climates, including predictions of different possible scenarios for the future.

Flavin, Christopher., and Nicholas Lenssen. *Power Surge: Guide to the Coming Energy Revolution*. New York: W. W. Norton, 1994. A thorough assessment by the Worldwatch Institute on the future potentials of renewable energy sources such as solar energy, wind energy, and biomass energy.

Fortey, Richard. *Earth: An Intimate History*. New York: Knopf, 2004. An elegant account by the renowned paleontologist of the history of plate motions and interactions, including discussions of the pioneers of geology who made these discoveries.

Fradkin, Philip L., *Magnitude 8: Earthquakes and Life Along the San Andreas Fault*. New York: Henry Holt & Company, 1998. A personal and historical account of the San Andreas Fault, combining the science of the fault with the history of the region.

Gillieson, David. *Caves: Processes, Development, Management*. Oxford: Blackwell Publishers, 1996. A discussion of cave evolution and development, emphasizing the delicate nature of cave ecosystems.

Gore, Al. *Earth in the Balance: Ecology and the Human Spirit*. New York: The Penguin Group, 1993. An early call of warning about the future directions of climate change, this book contains an excellent survey of the past effects of climate change on human history.

———. *An Inconvenient Truth*. New York: Rodale Books, 2006. A beautiful combination of photos, figures, and text that is based upon the excellent and influential movie of the same name.

Goudie, Andrew S. *Great Warm Deserts of the World*. New York: Oxford University Press, 2003. An account, region by region, of the world's great warm deserts, describing the geologic and climatic forces that have created them.

Gould, Stephen J. *Time's Arrow, Time's Cycle.* Cambridge, MA: Harvard University Press, 1988. A fascinating examination of the different perceptions of time and the age of Earth, particularly showing the changing and competing views of geologists.

Gross, M. Grant, and Elizabeth Gross. *Oceanography: A View of the Earth.* Upper Saddle River, NJ: Prentice Hall, 1995. A popular introductory text to oceans' complex systems and their impact on our lives and futures.

Gurnis, Michael. "Sculpting the Earth from Inside Out." *Scientific American,* (March 2001): 40–47. One of the world's top geodynamicists explains for a lay audience the connections between mantle convection and tectonic plate motions.

Hambrey, Michael, and Jurg Alean. *Glaciers.* Cambridge: Cambridge University Press, 2004. A nice montage of glaciers, with different themes related to glaciers presented along with the authors' own photographs.

Harris, Stephen L. *Fire Mountains of the West: The Cascade and Mono Lake Volcanoes.* Missoula, MT: Mountain Press Publishing Company, 1988. A great review of the many volcanoes of the Pacific Northwest and an assessment of their eruptive potential.

Hawking, Stephen. *A Brief History of Time.* Toronto: Bantam Books, 1988. The classic discussion of how our universe formed that set the standard for all other discussions of cosmology.

Hinrichs, Roger A., and Merlin Kleinbach. *Energy, its Use and the Environment.* Burgin, KY: Thomson Brooks/Cole Florence, 2005. A well-written college-level text on the many different issues related to energy use from a variety of different sources.

Hoffman, Paul F., and Daniel Schrag, "Snowball Earth." *Scientific American,* 282 (January 2000): 68–75. A description of one of the most fascinating and controversial theories in climate history: the idea that Earth went through a period of repeated deep-freezes that even caused the surfaces of the oceans to freeze.

Hough, Susan. *Earthshaking Science.* Princeton, NJ: Princeton University Press, 2004. An excellently written assessment of earthquake hazards and how they are determined.

Jacobson, Michael, Robert Charlson, Henning Rodhe, and Gordon Oriens. *Earth System Science From Biogeochemical Cycles to Global Changes.* New York: Academic Press, 2000. Earth systems science is the approach to earth science that emerged in the 1990s as an integrated, process-oriented means of study; this book explains how it works and how humans are interconnected with it.

Karato, Shun-Ichiro. *The Dynamic Structure of the Deep Earth.* Princeton, NJ: Princeton University Press, 2003. If you want to dig deeply, this advanced text by one of the world's leading experts provides descriptions of how and why Earth deforms.

Kasting, James F. "The Origins of Water on Earth." *Scientific American* 13.3 (2003): 28–33. An excellent assessment of where our water came from, where it is, and where it is going.

Keary, Philip, and Fred Vine. *Global Tectonics.* Oxford: Blackwell Scientific, 1990. A classic text that focuses on the way the ocean seafloor reveals the motions of the plates.

Keller, Edward A., and Robert Blodgett. *Natural Hazards: Earth's Processes as Hazards, Disasters, and Catastrophes.* Upper Saddle River, NJ: Prentice Hall, 2005. A very good overview of the many and varied types of natural disasters.

Keller, Edward A., and Nicholas Pinter, *Active Tectonics.* Upper Saddle River, NJ: Prentice Hall, 1995. A well-written discussion of folding and faulting and their relationships to plate tectonics, earthquakes, and mountain building.

Kuhn, Thomas S. *The Structure of Scientific Revolutions.* Chicago: University of Chicago Press, 1970. This landmark text on the philosophy of science highlights plate tectonics as an ideal example of a scientific paradigm shift in understanding.

Logan, William B. *Dirt.* New York: W. W. Norton and Company, 2006. A very personal and elegantly written account of a topic (soil) that does not often get a lot of respect.

Longview Publishing Company Staff. *Volcano, The Eruption of Mount St. Helens.* Seattle: Madrona Publishing, 1980. Excellent work by the Pulitzer Prize winning staff of *The Daily News* (Longview, Washington) with the assistance of the *Journal American* staff (Bellevue, Washington).

Lopes, Rosaly. *The Volcano Adventure Guide.* Cambridge: Cambridge University Press, 2005. This book is designed to prepare people to visit a volcano, giving logistical information on visiting 42 different volcanoes.

Macdougall, Douglas. *Frozen Earth: The Once and Future Story of Ice Ages.* San Diego: University of California Press, 2004. An exciting description of the last Ice Age and the significant ways it affected human and other biological life as it melted and receded.

McGregor, Glenn, and Simon Nieuwolt. *Tropical Climatology: An Introduction to the Climates of the Low Latitudes.* New York: John Wiley & Sons, 1998. This book provides a geographical view of physical process in the tropical atmosphere, offers explanations of how a location's climate is a product of these processes, and highlights the implications of tropical atmosphere behavior and climate change.

McPhee, John. *Assembling California.* New York: Farrar, Straus and Giroux, 1994. An intelligent discussion of the complex geologic history of California. Do not miss the stunning final chapter, which traces the damage of the advancing seismic waves from the 1989 Loma Prieta earthquake second by second.

————. "Atchafalaya" in *The Control of Nature.* New York: Farrar, Straus and Giroux, 1989: 3–92. A startling account of the magnitude of the task involved with keeping the Mississippi River in its course and of the inevitability of its eventual escape.

————. *Basin and Range.* New York: Farrar, Straus and Giroux, 1981. The classic story of how the West was formed, both geologically and culturally.

————. "Cooling the Lava" in *The Control of Nature.* New York: Farrar, Straus and Giroux, 1989: 95–179. An elegantly written account of the attempts to stop the flow of lava at two hot spot locations: Iceland and Hawaii.

————. "Los Angeles Against the Mountains" in *The Control of Nature.* New York: Farrar, Straus and Giroux, 1989: 183–272. A fun and entertaining account of Los Angeles' attempts to control the debris flows that continue to damage houses built upon the slopes of the San Gabriel Mountains.

————. *Rising from the Plains.* New York: Farrar, Straus and Giroux, 1986. An engrossing account of the geology of Wyoming, based upon a drive across the state, that weaves in the history of the settlement of this region.

Menard, H. William. *The Ocean of Truth.* Princeton, NJ: Princeton University Press, 1986. A retelling of the discovery of plate tectonics through seafloor observations by one of the early discoverers that makes you feel like you were right there as it happened.

Miller, Russell. *Continents in Collision.* Alexandria, VA: Time-Life Books, 1983. A nicely illustrated popular book that provides the fundamentals of plate tectonics.

Mithen, Steven. *After the Ice: A Global Human History 20,000–5000 BC.* Cambridge, MA: Harvard University Press, 2006. Reading at times like a time-travel novel, this fascinating portrayal of the end of the Ice Ages provides a wonderful sense of what the world was like as the ice was melting and civilization was starting.

Moran, Joseph, and Michael Morgan. *Meteorology: The Atmosphere and the Science of Weathering.* Upper Saddle River, NJ: Prentice Hall, 1996. An introductory text on the atmospheric aspects of environmental concerns with the basics of meteorology and climatology.

Morrison, David, and Tobias Owen. *The Planetary System.* San Francisco: Benjamin Cummings, 2002. A widely used general text on planetary astronomy that includes perspectives on the study of the origin, evolution, and distribution of life within our solar system and other planetary systems.

National Research Council Committee on the Alaska Earthquake. *The Great Alaska Earthquake of 1964.* Washington, DC: National Academy of Sciences, 1973. Accounts and descriptions of the largest earthquake known to have occurred in the United States.

Nicolas, Adolphe. *The Mid-Oceanic Ridges: Mountains Below Sea Level.* Berlin: Springer Verlag, 1995. A good college-level text on the structure and dynamics of the divergent boundaries between plates.

Officer, Charles B., and Jake Page. *Tales of the Earth.* New York: Oxford University Press, 1994. A nice selection of essays by a geophysicist (Officer) and a science writer (Page) that highlight some of the most interesting stories about the impact of geology on humanity.

Oldroyd, David. *Thinking about the Earth: A History of Ideas in Geology.* Cambridge, MA: Harvard University Press, 1996. A comprehensive overview of how people from ancient times to the present have tried to understand Earth.

Palin, Michael. *Himalaya*. London: Weidenfeld Nicolson Illustrated, 2004. Based on the BBC television series, this gives a wonderful account of a six-month trek around the Himalayan mountain range.

Pearce, Fred. *Deep Jungle*. Cornwall, England: Eden Books, 2005. A fascinating account of jungles, from their early exploration to modern scientific assessments of their diversity and complexity.

Pellant, Helen, ed. *Rocks and Minerals*. New York: DK Publishing, 2002. An attractive presentation of the many and varied types of minerals.

Pimm, Stuart. *The World According to Pimm: A Scientist Audits the Earth*. New York: McGraw-Hill, 2001. An assessment of human impact on Earth's surface, maintaining a sense of optimism in the face of often disturbing data.

Post, Austin, and Edward R. Lachapelle. *Glacier Ice*. Toronto: University of Toronto Press, 2000. The book features aerial and land-based photographs of North American glaciers, with an introductory explanation of glaciology.

Powell, James L. *Grand Canyon: Solving the Earth's Grandest Puzzle*. New York: Pi Press, 2005. An account of the geology of the world's most dramatic stream-erosional feature: the Grand Canyon.

Prager, Ellen, and Sylvia Earle. *The Oceans*. New York: McGraw-Hill, 2001. A wonderful account of the oceans, written to be both entertaining and educational.

Pugh, David. *Changing Sea Levels*. Cambridge: Cambridge University Press, 2004. A thorough discussion of the changes in sea level over short and long time scales, their effects on local geology and biology, and their implications for human society.

Robinson, Kim Stanley. *Red Mars*. New York: Bantam Spectra, 1993. (Also *Green Mars* and *Blue Mars*.) Science fiction, but you would not know it—this trilogy predicts a possible colonization of Mars in remarkably believable ways.

Rosenfeld, Charles, and Robert Cooke. *Earthfire: The Eruption of Mount St. Helens*. Cambridge, MA: MIT Press, 1982. A description of the events of the violent 1980 eruption, connecting them with the geologic processes that caused it.

Savoy, Lauret E., Eldridge Moores, and Judith Moores, ed. *Bedrock: Writers on the Wonders of Geology*. San Antonio, TX: Trinity University Press, 2006. A fascinating collection of writings, both fiction and non-fiction, that deal with the many aspects of geology.

Schopf, J. William. *Cradle of Life*. Princeton, NJ: Princeton University Press, 2001. A firsthand account of the discovery of some of Earth's earliest fossils and a discussion of what the implications are for the evolution of life in the universe.

Sigurdsson, Haraldur. *Encyclopedia of Volcanoes.*: San Diego, CA: Elsevier Science and Technology Books, 1999. Everything that you ever wanted to know about volcanoes, with plenty of opportunity to ask in these 1,359 pages!

————. *Melting the Earth*. New York: Oxford University Press, 2006. A well-written account of the history of volcanology, showing how our perceptions of melting, magma, and volcanoes have changed over time.

Smith, Keith, and Roy Ward. *Floods: Physical Processes and Human Impacts*. New York: John Wiley & Sons, 1998. A good overview of floods and their impacts, hazards, and assessments, using case examples from historic floods.

Sobel, Dava. *The Planets*. New York: Penguin Books, 2006. A graceful and elegant discussion of the solar system that weaves the science together with popular culture, mythology, and science fiction.

Stacey, Frank D. *Physics of the Earth*, 3rd ed. Brisbane, Australia: Brookfield Press, 1997. Required reading for all geophysicists, this book provides tremendous insights but requires a good foundation in math and physics.

Stein, Seth, and Michael Wysession. *Introduction to Seismology, Earthquakes, and Earth Structure*. Oxford: Blackwell Scientific, 2003. The leading upper-level undergraduate and beginning graduate-level text on earthquakes and seismic waves. It provides the means of imaging our planet's interior, just in case you want to dig deeper.

Sullivan, Walter. *Continents in Motion*. New York: McGraw-Hill, 1974. Walter was the long-time editor of the New York Times' *Science Times*; here he presents the discovery of plate tectonics as it passed across his desk.

Tarbuck, Edward J., and Fred Lutgens. *The Atmosphere: An Introduction to Meteorology*, *10th* ed. Upper Saddle River, NJ: Prentice Hall, 2007. An excellent, solidly written college textbook on weather and the atmosphere.

———. *Earth: An Introduction to Physical Geology*, *9th* ed., Upper Saddle River, NJ: Prentice Hall, 2008. The most widely used college introductory geology textbook: solid, straightforward, well-written, up-to-date, and with excellent illustrations.

Trujillo, Alan P., and Harold Thurman. *Essentials of Oceanography*, *9th* ed. Upper Saddle River, NJ: Prentice Hall, 2008. An excellent college-level textbook that provides the fundamentals of the study of the oceans.

Tyson, Neil deGrasse. *Death by Black Hole: And Other Cosmic Quandaries.* New York: Norton, 2007. An excellently written collection of essays on cosmology from the world's best-known current spokesman for astronomy and astrophysics.

Uyeda, Seiya. *The New View of the Earth.* New York: W.H. Freeman & Company, 1995. A famous text that gives a very insightful description of the how and why of plate tectonics.

Van Andel, Tjeerd H. *New Views on an Old Planet: Continental Drift and the History of the Earth.* Cambridge: Cambridge University Press, 1994. A classic book, widely read, that weaves together the evolution of the planet, the history of the oceans and atmosphere, and the evolution of life.

Vogel, Shawna. *Naked Earth.* New York: Dutton Adult, 1995. An excellent science writer describes the modern understanding of Earth's interior in highly readable and accessible text.

Vonnegut, Kurt. *Cat's Cradle.* New York: Dell, 1971. A fictional account of the end of the world where a high-pressure form of water ("Ice 9") plays a pivotal role.

Walker, Gabrielle. *An Ocean of Air.* Orlando, FL: Harcourt, 2007. An entertaining account of the scientific study of our atmosphere and the air it contains.

Ward, Peter, and Donald Brownlee. *Rare Earth.* New York: Springer-Verlag, 2003. An extremely influential book that demonstrated that the conditions required for continuous and stable life on the surface of a planet (needed for the evolution of complex life forms) may be exceedingly rare, and that there might not be many planets like Earth in our galaxy.

Webb, Stephen. *If the Universe Is Teeming with Aliens … Where Is Everybody?: Fifty Solutions to the Fermi Paradox and the Problem of Extraterrestrial Life*. New York: Copernicus Books, 2002. Fifty different possible solutions to Fermi's Paradox (Fermi is claimed to have asked, following a discussion of phenomena like flying saucers in 1950, "Where is everybody?"), including the author's favorite—that life may be ubiquitous throughout the galaxy, but the conditions required for the evolution of complex life might be extremely rare.

Westbroek, Peter. *Life as a Geological Force: Dynamics of the Earth*. New York: W. W. Norton & Co., 1992. An interesting presentation of the ways that the geosphere and biosphere are interrelated, each significantly altering the other.

Winchester, Simon. *A Crack in the Edge of the World: America and the Great California Earthquake of 1906*. New York: Harper Collins, 2005. A thorough account of the great earthquake of 1906 and the geology of California and the San Andreas Fault.

———. *Krakatoa*. New York: Harper Perennial, 2005. A detailed description of the geology of Indonesia and the historical events concerning the great eruption of 1883.

Yergin, Daniel. *The Prize: The Epic Quest for Oil, Money and Power*. New York: Simon & Schuster, 1993. The well-known energy consultant provides his assessment of the political, economic, cultural, and environmental issues surrounding petroleum as an energy source.

Zigler, Alan. *Hawaiian Natural History, Ecology, and Evolution*. Honolulu: University of Hawaii Press, 2002. This book traces (with words and pictures) the natural history of Hawaii through such topics as island formation, plant and animal evolution, and the effects of humans and exotic animals on the environment.

Recommended Websites:

http://www.nasa.gov/. The website for NASA (National Aeronautics and Space Administration) has an enormous amount of information, including photographs from every single NASA mission. There are great educational materials and lots of features and stories, but you will be amazed by the number of pictures all those NASA satellites, rovers, and astronauts have taken over the years!

http://www.noaa.gov/. The website of the National Oceanic and Atmospheric Administration (NOAA) is the place to go for all things related to the oceans and atmospheres. As well as having feature stories and materials for teachers, it has an up-to-date weather watch that graphically displays all the problematic weather-related events around the world (e.g., floods, hurricanes, droughts, and forest fires).

http://www.usgs.gov/. The website of the United States Geologic Survey (USGS) has a tremendous amount of information in the areas of geology, geography, hydrology, and biology. It has maps, educational activities, regional science topics, earth science trivia, and a regularly updated natural hazards section that describes the latest earthquakes and volcanoes.

Notes

Notes